FIRE
—AND—
EXPLOSION
PROTECTION SYSTEMS

A Design Professional's Introduction

Second Edition

Michael R. Lindeburg, PE

The Power to Pass
www.ppi2pass.com

Professional Publications, Inc. • Belmont, California

Benefit by Registering This Book with PPI

- Get book updates and corrections
- Hear the latest exam news
- Obtain exclusive exam tips and strategies
- Receive special discounts

Register your book at **www.ppi2pass.com/register**.

Report Errors and View Corrections for This Book

PPI is grateful to every reader who notifies us of a possible error. Your feedback allows us to improve the quality and accuracy of our products. You can report errata and view corrections at **www.ppi2pass.com/errata**.

FIRE AND EXPLOSION PROTECTION SYSTEMS
Second Edition

Current printing of this edition: 6

Printing History

edition number	printing number	update
2	4	Minor corrections.
2	5	Minor corrections. Copyright update.
2	6	Minor corrections.

Printed in the United States of America

PPI
1250 Fifth Avenue, Belmont, CA 94002
(650) 593-9119
www.ppi2pass.com

Library of Congress Cataloging-in-Publication Data

Lindeburg, Michael R.
    Fire and explosion protection systems : a design professional's
introduction / Michael R. Lindeburg. -- 2nd ed.
        p.   cm. -- (Engineering reference manual series)
    Includes index.
    ISBN: 978-0-912045-82-5
    1. Fire prevention--Equipment and supplies   2. Fire extinction--Equipment and supplies.   3. Explosions.   I. Title.   II. Series.
TH9245.L56   1995
693'.82--dc20   94-23456
                                                                       CIP

TABLE OF CONTENTS

PART 1: Building and Plant Components

PART 2: Detection and Warning Systems

PART 3: Fire Protection Methods

PART 4: Design of Water Sprinkler Systems

PROFESSIONAL PUBLICATIONS, INC.

PART 5: Explosion Protection Systems

PART 6: Practice Problems

Appendices

Index

PREFACE and ACKNOWLEDGMENTS
to the Second Edition

The main difference between this edition of *Fire and Explosion Protection Systems* and the first edition is the set of ten solved practice problems (with solutions) that has been added. A few other minor items of specific importance have also been added to the text.

Mr. Kenneth J. Harris, PE (Fair Oaks, CA), was the major contributor to this second edition. Once the two of us had agreed on the subject areas that lent themselves to additional practice with numerical examples, he wrote and solved problems in those subjects. I suspect that most readers will learn their fire and explosion protection material primarily from solving the problems he wrote.

Robert G. Purington, fire and safety consultant (Arnold, CA), was the developmental and technical editor who oversaw the validity of the problems and their solutions.

I continue to appreciate the outstanding work of the acquisitions and production staffs at PPI. Gerald R. Galbo, acquisitions editor, coordinated the gathering and checking of the new material for this edition. Jessica R. Whitney-Holden, managing production editor, oversaw the production of the new edition. Mia Laurence copyedited the new material, Sylvia M. Osias did the typesetting, and Charles P. Oey did the illustrations. This edition owes its quality appearance to them.

Thanks to you all!

Michael R. Lindeburg, PE

PREFACE to the First Edition

I wrote this publication to serve primarily as a review for any fire protection question on the National Council of Examiners for Engineering and Surveying (NCEES) PE examination for mechanical engineers.

The subject of fire protection has been part of the mechanical PE examination for many years. But until now, no single exam-oriented publication has consolidated the subjects tested on the examination. Part of the difficulty in writing a publication such as this is that fire protection covers a large body of knowledge. Some of this knowledge (e.g., flow of water through sprinkler pipes) is common to traditional engineering, some is empirical, and some is dictated by code.

The absence of an adequate exam-oriented reference has meant that mechanical engineers have not been able to learn this subject. This publication changes that.

In reviewing the NFPA source material during the outline stage, I realized that the scope of this publication would have to be limited. For this edition, at least, I made several assumptions about what the average mechanical engineer would generally need to know for an effective exam review, and I excluded all other material. This means that I have only covered a sliver from the entire log of fire protection knowledge.

For example, to be an effective professional, a fire protection engineer must have knowledge of fire dynamics, be able to perform computer hazard simulations, be able to calculate fire resistance, and know how to design smoke control systems. I did not consider these to be testable subjects, so I omitted them.

I emphasized commercial fire protection systems. Residential systems are mentioned in passing, but this publication does not focus on special provisions pertaining to residential buildings. I reasoned that residential designs are completed by architects and sprinkler contractors, not mechanical engineers.

And, as this is a publication for mechanical engineers, I resisted getting too structural. This means that building subjects in ASCE's *Handbook of Structural Fire Protection* were not included in this publication.

Similarly, I resisted including any information about special sprinklers. Design of large drop and early suppression fast-response (ESFR) sprinklers is not covered. Including data on these sprinklers would not change the theory presented in this publication, but it would require including almost all of *Standard for the Installation of Sprinkler Systems* (NFPA 13), the primary source of testable information.

Out-of-the-ordinary equipment and applications that, in real life, would require tests to determine suitability for use are not covered.

I omitted building code subjects that I thought were in the domains of architects and interior designers. There are no discussions of maximum occupancies, design of firewalls, maximum and minimum distances between windows and doors, or flammability of carpets and furniture. These are important subjects, but they are not normally dealt with by mechanical engineers.

I only covered a few specific installation types (e.g., power-generating plants), and even at that, the coverage is general. I avoided discussing other special occupancies, which means you will have to go elsewhere to learn how to protect aircraft hangars, missile assembly sites, rubber tire warehouses, and the like.

No one, not even an expert in the field of fire protection, can be knowledgeable in all of the NFPA specifications and standards—there are just too many of them. I have worked primarily from *Fire Prevention Code* (NFPA 1), *Standard for the Installation of Sprinkler Systems* (NFPA 13), and *Centrifugal Fire Pumps* (NFPA 20). I have done my best to include general, testable information from these specifications.

Use of metric units is one of the issues that I probably will not address to everyone's satisfaction. Of course, the traditional units of pounds per square inch (psi) and gallons per minute (gpm) are included. However, most modern fire protection design is done in SI units. I, too, have a policy of presenting everything in SI units. Interestingly, though, one of the primary reference for this publication (*Standard for the Installation of Sprinkler Systems*, NFPA 13) uses bars to measure pressure, so this nonstandard metric unit is also included.

This publication is ethical in its authorship. In writing it, I did not use exam problems, problem reconstructions, or look-alike problems. I have not released or used information that would lead you toward studying specific problems on future exams. All information used in determining the scope and depth of coverage was obtained by legitimate and legal means.

I hope this publication will serve you well in your career.

Michael R. Lindeburg, PE

ACKNOWLEDGMENTS

I wish to acknowledge the help given me by Peter Lazdins (Compufire Design Services, Clearwater, FL) and Jonathan R. Barnett, Ph.D. (Center for Firesafety Studies, Worcester Polytechnic Institute, Worcester, MA), who reviewed the original draft of this material. Their knowledge of the theoretical and familiarity with the practical took many of the rough edges off the material. Without their comments, this publication would have had much more of an academic flavor.

Martin Jay Hanna III, P.E. (Hanna Fire Engineering Corporation, Baltimore, MD), served as developmental editor for this publication. He reviewed the manuscript after I rewrote and expanded the material following the initial reviews. You will never know the areas that were improved by his involvement, but you will surely reap the benefits of it.

All of these reviewers made suggestions that contributed to the usefulness of this publication. However, I alone am responsible for any of the inevitable errors, inconsistencies, and contradictions that are discovered.

I again offer my thanks to the Production Department at PPI. Wendy Nelson was my in-house editor. She arranged the critiques and reviews that improved the manuscript. Kurt Stephan served as copy editor. Jenny Pasqual did the illustrations. Mary Christensson obtained the permissions. Sylvia Osias did the typesetting. Jessica Whitney-Holden proofed the galley and page proofs. The level of professionalism exhibited by this team, under the leadership of its manager, Lisa Rominger, is truly phenomenal. The degrees of accuracy and perfectionism that this publication enjoys meet and exceed those of the best technical publications.

Thanks to you all!

Michael R. Lindeburg, PE

LIST OF TABLES

NOMENCLATURE

symbol	term	units U.S.	SI
A	area	ft^2	m^2
A	fraction of air in tank	–	–
C	Hazen-Williams coefficient	–	–
C	vapor concentration	–	–
d	internal pipe diameter	in	mm
g	acceleration due to gravity	ft/sec^2	m/s^2
h	height of fluid column	ft	m
k	mixing efficiency factor	–	–
K	orifice constant	gal/min-$\sqrt{\text{psi}}$	ℓ/s$\sqrt{\text{kPa}}$
L	length or distance	ft	m
m	mass	lbm	kg
$\dot{m}$	mass leak rate	lbm/sec	kg/s
MW	molecular weight	lbm/lbmole	kg/kmol
n	number of air changes	–	–
n	number of sprinklers	–	–
p	absolute pressure	psfa	kPa
p	gauge pressure	psig	kPa
q	water flow rate increment	gal/min	ℓ/s
Q	pump discharge rate	gal/min	ℓ/s
Q	total location water-flow rate	gal/min	ℓ/s
Q	calculated air introduction rate	ft^3/min	ℓ/s
R	specific gas constant	ft-lbf/lbm-°R	kJ/kg·K
R^*	universal gas constant	ft-lbf/lbmole-°R	kJ/kmol·K
S	distance	ft	m
t	time	sec	s
T	absolute temperature	°R	K
v	velocity	ft/sec	m/s
V	volume	gal	ℓ
$\dot{V}$	volumetric flow rate	gal/sec	ℓ/s
γ	specific weight	lbf/ft^3	n.a.
ρ	mass density	lbm/ft^3	kg/m^3

subscripts

d	discharge	n	normal
e	elevation difference	t	total
f	friction	v	velocity

PART 1:
Building and Plant Components

1. Fire and Explosion Protection

Design of fire and explosion protection systems encompasses the design of buildings, fire detection systems, alarm and communications systems, fire suppression systems, and smoke management systems. Emphasis during the design phase extends far beyond the design of the monitoring and alarm functions, focusing also on hazard- and damage-minimization techniques.

2. Building Codes

The design, installation, and operation of fire detection, alarm, and suppression systems are governed by a blend of standard engineering techniques, local and state codes, and fire insurance requirements. It is necessary to review all applicable codes and meet the most stringent of these overlapping requirements. Furthermore, although codes mandate minimum standards, engineering analysis, common sense, and professional judgment should be used in specifying the level of fire protection in a building.

Design of buildings and installation of fire protection systems are governed by local building codes. A municipality can develop its own building code, but it usually adopts one of the three model (standard) codes: the *Uniform Building Code* (International Congress of Building Officials, ICBO), the *Basic Building Code* (Building Officials and Code Administrators, BOCA), and the *Standard Building Code* (Southern Building Code Congress International, SBCCI).

Most of these model building codes have been heavily influenced by, or have adopted sections largely unchanged from, National Fire Protection Association (NFPA) publications such as the *NFPA Fire Prevention Code* (NFPA 1), *Life Safety Code* (NFPA 101), and the *Uniform Fire Code*, published by the ICBO in conjunction with the Western Fire Chiefs Association.

Additional requirements for handicapped persons may be imposed by the Occupational Safety and Health Administration (OSHA), the Department of Housing and Urban Development (HUD), and the Department of Health, Education, and Welfare (HEW), as well as by federal and state statutes that might be passed from time to time, such as the Americans with Disabilities Act (ADA).

Most engineers are unfamiliar with the code-related aspects of designing for fire-protection. This is to be expected. The requirements vary from city to city, depending on which building code has been adopted. Furthermore, regulations concerning sprinklers and other protective devices are subject to change, administrative review, and interpretation by local building and fire officials. For these reasons, architects and fire protection engineers with specialty knowledge and building and fire officials should be consulted early in the planning stages of any new project.

3. Occupancy Hazard Type

The type of occupancy affects the required degree of protection. Nursing homes, schools, hospitals, and office buildings have greatly different needs. Similarly, multistory buildings with limited escape routes demand greater fire protection than single-level dwellings.

Standard for the Installation of Sprinkler Systems (NFPA 13) specifies three types of relative fire hazards for buildings: light, ordinary, and extra hazard. The occupancy classification determines many aspects of the fire protection design, including (in the case of sprinklers) required water pressure, flow rate, and supply duration (see Sec. 52).

It should be noted, however, that the occupancy type for the purpose of sprinkler design may not always coincide with the occupancy for hazard classification. For example, an office building, normally classified as a light hazard occupancy, may contain a large amount of combustible material (such as papers and files), warranting an upgrade to ordinary hazard occupancy in fire protection occupancy classification.

Areas of *light hazard occupancy* have low quantities of combustible material and experience fires with low heat-generation rates. There is generally no processing, manufacturing, or large scale storage. No flammable or combustible liquids are stored. Examples of light hazard occupancies (as determined by NFPA 13) are churches, clubs, educational institutions, hospitals, libraries (except large stack rooms), museums, nursing and convalescent homes, offices (including data processing facilities), residences, restaurant seating areas, theaters, and auditoriums (excluding stages and unused attics).

Ordinary hazard occupancies are divided into two sub-classifications. *Ordinary hazard group 1* occupancies have low combustibility, moderate levels of combustibles, and stockpiles of combustibles less than 8 feet (2.4 m) high, and they experience fires with moderate heat release rates. Flammable and combustible liquids and gases are limited in quantity or are placed in the most fire-safe locations. This occupancy has the least stringent sprinkler design requirements of the ordinary hazard occupancies. Examples are parking garages, bakeries, beverage plants, canneries, dairy product manufacturing and processing plants, electronic plants, laundries, and restaurant service areas.

Ordinary hazard group 2 classifications represent average conditions for manufacturing and processing industries. These include areas with contents in moderate quantities and with moderate to high combustibilities, stockpiles that do not exceed 12 feet (3.7 m) in height, and moderate to high heat-release rates. Examples are cereal mills, ordinary chemical plants, cold storage warehouses, distilleries, libraries with large stacks, machine shops, mercantile businesses, metal working plants, printing and publishing plants, and locations where wood products are assembled.

The ordinary hazard group 2 classification also applies to areas containing large quantities of highly combustible contents and where heat generation rate will be high. Examples include feed mills, paper and pulp mills, paper processing plants, piers and wharves, repair garages, tire manufacturing plants, warehouses (containing paper, household furniture, paint, liquor, etc.), and wood machining operations.

Extra hazard occupancies contain large quantities of highly combustible materials (e.g., flammable and combustible liquids, dusts, etc.). The *extra hazard group 1*

classification includes occupancies with little or no flammable or combustible liquids, while *extra hazard group 2* occupancies include all areas with moderate to substantial amounts of such liquids.

Extra hazard group 1 occupancies include users of hydraulic fluid under pressure (which, when lines are ruptured, will produce fine sprays of oil), die-casting and metal-extruding factories, printing plants, rubber manufacturing and processing plants, saw mills, textile mills, and factories handling foam rubber.

Extra hazard group 2 occupancies include users of asphalt and manufacturing operations where oil, cleaning solvents, varnishes, paint, and other finishes are used openly. Extra hazard group 2 also includes locations where water sprinkler discharge is likely to be obstructed by nonstructural conditions.

An additional hazard classification is the *high-piled storage system* classification. This is primarily a description of warehouses with storage in the form of solid piled, palletized, rack storage, bin box, and shelf storage higher than 12 feet (3.7 m). A different approach is taken in designing fire protection systems for high-piled storage areas.

There are many other occupancy types that are determined by use, and these occupancies may be governed by standards far different from the general standards in this publication. Specific examples and their corresponding NFPA standards are: flammable liquids storage locations (*Flammable Liquids Code*, NFPA 30), aerosol liquids storage locations (*Aerosol Products*, NFPA 30B), general indoor warehouse storage (*General Storage*, NFPA 231), and rubber tire storage locations (*Storage of Rubber Tires*, NFPA 231D), just to name a few.

4. Building Fire Resistance Classifications

Buildings and other structures are also classified according to the effect fire will have on them. There is no classification known as *fireproof*, although the term has been used in some codes in the past.

Combustible structures are constructed of combustible materials in such a manner that significant damage and/or total destruction will result from a fire.

Heavy timber construction incorporates masonry exterior walls and interiors that are supported by heavy timber members.[1] The key concept in determining the

---

[1]The term *timber* generally means lumber more than 5 inches (14 cm) in the smallest dimension. In the past, timbers has been a subcategory of lumber. The current trend is toward using *wood members* to refer to all sizes of building material.

classification is that massive timber members are slow to ignite, and that once ignited, they burn slowly, delaying the collapse.

Ordinary construction consists of masonry walls with wooden floor and roof diaphragms. Such buildings are quick-burning and are expected to collapse in a fire and, hence, are sometimes referred to as having *quick-burning* construction. Their primary benefit is in a reduction in potential for conflagration.

Wood frame (i.e., *stick*) *construction* consists of wood construction throughout. The addition of a brick veneer does not change the classification.

Structures with *noncombustible construction* will not readily burn but do not qualify for fire-resistive construction classification. Noncombustible structural members may, nevertheless, be damaged in a fire. For example, unprotected steel girders and columns will not burn, but they lose their strength, elongate, and transfer heat in a fire. Steel is protected in a number of ways (see Sec. 7).

Fire-resistive construction, such as structures incorporating concrete and protected steel members, will not easily burn and possesses the ability to withstand short-term fires without significant damage. (The level of fire resistance of concrete members is increased by increasing the thickness of the concrete cover over the reinforcing steel.) Fire-resistive assemblies are not necessarily noncombustible. For examples, floors and walls of wood and gypsum board are fire resistive, although they are combustible.

Table 1 compares the building classifications by model building code. (Table 1 is not intended to indicate that different codes have identical requirements for any specific type of construction.)

5. Building Fuel Loading

Code regulations governing the construction (i.e., limiting the size and occupancy, specifying features, etc.) are based on the expected fire risk. Such risk is, to a large extent, determined by the *fuel loading* (fire loading) in the building. Fuel loading is expressed in pounds of equivalent combustible material per square foot of floor area (kg/m^2).

All loading is expressed in terms of ordinary combustibles, such as wood, which has a heating value of approximately 8000 BTU/lbm (4400 kcal/kg). The fuel values of building contents are included for all types of buildings. For combustible buildings, the fuel content of the construction materials is also included.

Table 1
Comparison of Building Classifications

construction type	BBC (BOCA)	SBC (SBCCI)	UBC (ICBO)	NFPA 220
fire resistive (FR)	1A	I	I (FR)	I (444)
	1B	II	I (FR)	I (332)
noncombustible protected	2A		II (4 hr)	II (222)
	2B	IV (1 hr)	II (1 hr)	II (111)
noncombustible unprotected (N)	2C	IV	II (N)	II (000)
heavy timber (HT)	4	III	IV (HT)	IV (2HH)
ordinary protected	3A	V (1 hr)	III (1 hr)	III (211)
ordinary unprotected (N)	3B	V	III (N)	III (200)
wood frame protected	5A	VI (1 hr)	V (1 hr)	V (111)
wood frame unprotected(N)	5B	VI	V (N)	V (000)

Table 2 lists typical fuel values for various building uses. No typical values are given for buildings containing hazardous conditions. Risk factors for hazardous locations are based on factors other than fuel loading.

Table 2
Typical Fuel Loading

building type	fuel loading	
	lbm/ft^2	kg/m^2
assembly/meeting hall	5–10	25–50
classroom	5–10	25–50
industrial	10–35	50–170
institutional	3–10	15–50
library	10–40	50–200
mercantile	10–20	50–100
offices	5–10	25–50
residential	5–10	25–50
storage, business files	10–40	50–200
storage, other	10–100	50–500

6. Thermal Properties of Building Materials

The thermal properties of building materials are sometimes useful in determining the degree of hazard. For example, the high thermal conductivity of metals can aid in spreading a fire. Thermal properties used for designing fire and explosion protection systems are no different than the properties used in HVAC and heat transfer calculations. App. A provides a summary of properties for selected building materials.

Mechanical and thermal properties of building materials are extremely temperature-dependent. Use of average properties without consideration of the temperature dependency can result in unsafe designs.

7. Protecting Steel Construction

Structural steel, though noncombustible, loses approximately 40% of its yield strength in the temperature range of 1000–1300°F (540–700°C). Also, steel expands upon heating, which contributes to the collapse of walls in some construction types.

Lighter steel members will experience strength reductions sooner than heavier members. For example, a W10×49 (a beam width of 10 inches and weight of 49 pounds per foot) needs to absorb less energy than a heavier W10×60. Unfortunately, prices of steel members are directly related to their weights (i.e., are priced by the pound), so architects and design engineers generally specify the lightest possible members.

Structural steel members can be temporarily protected from losing strength in a fire by encasement, attaching coverings, spraying with fireproofing, using membrane protection, and by sprinklers. *Water Spray Fixed Systems* (NFPA 15) and *Rack Storage of Materials* (NFPA 231C) cover sprinkler protection of steel legs and other support members.

With *encasement*, each steel member is covered with a noncombustible material. Poured concrete is typically used, and the resulting concrete-steel member is known as a *composite structural member*. Covering with sheets of gypsum wallboard and rockwool (mineral wool) is also common. The high thermal capacity of concrete makes up for its relatively high thermal conductivity (which transmits heat to the protected steel member). Hydrated water is vaporized in both concrete and wallboard, and the heat of vaporization is effectively removed from the protected member.

Coverings of tile, concrete block, other heavy masonry, and lath and plaster are also occasionally encountered though no longer commonly specified.

Gypsum wallboard is available in two types—type X and regular. Type X contains inorganic fibers that increase the wallboard's strength. Gypsum is inexpensive and readily available. It is also 50–75% lighter than sprayed concrete coverings, is easy to apply, and leaves a regular finish that can be painted or incorporated into the interior decor. Gypsum wallboard can be attached by wire and wireboard screws, or thin-gauge sheet metal covers can be used.

Sprayed-on fireproofing is more common with larger structures. The steel members are sprayed with an inorganic (mineral) fiber, an intumescent, gypsum, or cement (e.g., shotcrete). In some cases, the fireproofing can be troweled on. Asbestos is no longer used as a component of mineral-based fireproofing material. The effectiveness of this method of protecting steel depends on obtaining adequate thickness and positive bonding between the steel and fireproofing. Cementious materials are generally preferred.

An *intumescent* is a paint-like coating containing mixtures of organic and inorganic substances. The coating, when exposed to a fire, swells to many times its original thickness and then chars. The layer of char forms an insulating blanket.

Magnesium oxychloride is another paint-like material that can be sprayed on steel members. Its basis for fire protection is the large (44–54% by weight) water of hydration content. The water is vaporized when the coating reaches approximately 570°F (300°C).

With *membrane protection*, entire diaphragms (e.g., floors and ceilings) or other large areas may be protected as a unit. For example, floors may be covered with a layer of poured concrete or gypcrete. Ceilings can be protected with fireproof ceiling tiles or panels.

8. Smoke Control

Smoke control is achieved by a combination of doors, dampers, and vents. All smoke control devices must be listed by UL or other approving agency.

Smoke doors are used to keep smoke and fire from spreading through passages intended for people. They are held open with floor- or wall-mounted electromagnets. All fire doors controlled by the monitoring system will close automatically on the first alarm. *Fire doors* are essentially the same as smoke doors, although their construction and purpose may be different.

All critical doors (e.g., stairwell exits) that are kept closed and locked for security reasons must incorporate a panic exit bar. The bar should override the door's lock, manually opening the door and sounding an alert signal.

Dampers are devices that automatically close or adjust to prevent the spread of smoke or fire through HVAC ducts and other passageways not traveled by occupants. *Fire dampers*, with a $1\frac{1}{2}$-hour fire rating, are constructed differently than smoke dampers. Fire dampers are required when an HVAC duct pierces a fire wall. They are almost always spring-loaded, relying on fusible links that melt upon exposure to heat. *Smoke dampers* may be motorized and triggered by a signal from a detector, or they may also be spring-loaded.

Motorized dampers must be capable of closing using the signal and power obtained from the fire panel's low-voltage backup battery.

In some installations, all supply and exhaust fans in the HVAC system are designed to shut down on the first alarm. However, in other installations, it is desirable to shut down only supply fans and to exhaust smoke from a building.

Roof vents are provided in roofs of buildings containing materials with high smoke- and heat-release potential such as warehouses, flammable liquid storage and handling facilities, and other extra-hazard occupancies. Roof vents are normally closed. They are opened, either manually or by intentional remote signal, to permit accumulated smoke and heat to be cleared from the building. *Smoke and Heat Venting Guide* (NFPA 204M) should be consulted for additional guidance.

Pressurization creates a positive air pressure to inhibit the influx of smoke from adjacent areas or floors above and below. In particular, stairwell escape routes out of high-rise buildings should be pressurized to provide a smoke-free exit path.

9. Ducts and Open Plenums

Ducts and plenums carry environmental air that must be kept free from smoke. (The term *environmental air* refers to the air that people breathe.) Generally, combustible material should be not be located in ducts and plenums. Rooms, hallways, and corridors should not be used for supply or return of environmental air. (Exceptions to these two restrictions should be carefully considered.)

Drop ceilings that use the space above as an open environmental air plenum create special problems and are treated as potential hazards in the National Electric Code and most fire codes. For example, electrical wiring run in ducts and plenums should normally be mineral-insulated metal-sheathed cable or else be run in electrical metallic tubing (EMT), rigid conduit or, in some cases, listed nonmetallic environmental airspace tubing (NEAT). Plastic pipe or conduit used in ducts and plenums must be approved or listed for that application.

Communication, computer, and other low-voltage wiring, including coaxial, multipaired, multiconductor, and fiber-optic varieties not run in metallic tubing, must be approved for such use. Teflon-coated cable is often approved for this use.

Smoke detectors located in drop ceiling plenums must be located properly, and they must include provisions for checking and maintenance. A remote alarm lamp should be brought down and mounted below the ceiling level to indicate which detector causes an alarm condition.

Smoke detectors used in ducts and plenums can be either of the photoelectric or ionization types. Smoke detectors located in ducts can be used to control smoke dampers and to prevent damage to HVAC equipment. However, due to the possibilities of inadequate or infrequent maintenance and improper location, duct detectors are not replacements for area smoke detectors (see Sec. 21). For these reasons, duct detectors are of specific, but limited, value.

As a general rule, duct smoke detectors are not required in air handling units carrying less than 15,000 cfm (425 m^3/min) to sprinklered buildings. *Air Conditioning and Ventilating Systems* (NFPA 90A) should be consulted for more information.

10. Elevator Capture

Some building codes require elevator cars to return to the first (or some designated alternate) floor in the event of a fire alarm, a technique known as *elevator capture*. The elevator control panel must be interconnected with the fire alarm panel. The low-voltage power produced by the fire panel's backup battery must be sufficient to communicate with the elevator controls.

Provisions must be made for fire fighters to override the elevator capture mechanism upon demand. When justified by professional judgment, fire fighters can then use the elevators to rescue trapped or nonambulatory occupants and reach burning areas.

Local codes should be studied in regard to elevator capture. Elevator capture is not allowed by the American National Standards Institute (ANSI) *Elevator Code*, which is used in some areas.

11. Volumetric Limits on Containers and Portable Tanks

In order to minimize the potential hazards of moving hazardous materials in containers and portable tanks, volumetric limits on contents have been established. Specific limits are given in *Flammable Liquids Code* (NFPA 30), and typical values are reproduced in Table 3. Other regulations will specify the safety features that containers and portable tanks must possess, the types of cabinets that must be used to store them, and maximum bulking (in terms of combined liquid volume and maximum pile height) of multiple containers.

Table 3
Maximum Allowable Size of Containers
and Portable Tanks

container type	flammable liquids			combustible liquids	
	Class IA	Class IB	Class IC	Class II	Class III
glass	1 pt	1 qt	1 gal	1 gal	5 gal
metal (other than DOT drums) or approved plastic	1 gal	5 gal	5 gal	5 gal	5 gal
safety cans	2 gal	5 gal	5 gal	5 gal	5 gal
metal drum (DOT spec.)	60 gal	60 gal	60 gal	60 gal	60 gal
approved metal portable tanks	660 gal	660 gal	660 gal	660 gal	660 gal
polyethylene DOT spec. 34, or as authorized by DOT exemption	1 gal	5 gal	5 gal	60 gal	60 gal

Multiply pints by 0.473 to obtain liters.
Multiply quarts by 0.95 to obtain liters.
Multiply gallons by 3.8 to obtain liters.

Reprinted with permission from *Flammable and Combustible Liquids Code Handbook*, copyright © 1991, National Fire Protection Association, Quincy, MA 02269.

12. Protecting Hydraulic Equipment

The primary danger in hydraulic systems using petroleum-based hydraulic fluids is leakage. Pressurized hydraulic fluid escaping as a fine mist through a pinhole or crack mixes with air and forms an easily ignitable mixture. Precautionary fire protection measures begin with minimizing leaks.

Tubing in hydraulic systems with pressures in excess of 200 psi (1380 kPa; 13.8 bars) should be SAE 1010 dead-soft, cold-drawn, seamless steel tubing or equivalent. A factor of safety of 8 based on the normal working pressure should be used.

Tubing is preferable to pipe in hydraulic systems. Solderless, steel fittings of the flareless locking sleeve-type or flare-type should be used. Threaded pipe should be avoided. When threaded connections are used, a safety factor of 8 based on the normal working pressure should be used.

Tubing runs should have as few bends as possible, though at least one bend is necessary to provide for thermal dimension changes. The minimum bend radius should be three tube diameters.

Flexible hose should be steel-reinforced and suitable for the hydraulic fluid being used. Its should have a pressure rating five times the actual operating pressure.

A well-marked emergency shutoff switch is required for each pump. The system should automatically switch off upon loss of hydraulic pressure and when activated by the sprinkler water-flow alarm, fusible link, or other fire detector.

Stationary hydraulic equipment housed in combustible construction or exceeding 100 gallons (380 ℓ) in aggregate capacity of hydraulic fluid should be protected by automatic sprinklers. Sprinklers should extend at least 20 feet (6.1 m) beyond the equipment. Sprinklers can be omitted for small systems if the construction is noncombustible, ignition sources are not normally present, and provisions exist for automatic or manual system shutdown.

To eliminate sources of ignition, all electrical equipment within 25 feet (7.6 m) of hydraulic pumps, meters, fittings, hoses, and lines should be listed for Class I, Division 2 use (see Secs. 94 and 95). Wiring should be in threaded, rigid, metal conduit.

13. Protecting Electrical- and Power-Generating Plants

Commercial power plants are prone to fires and explosions. These hazards exist during normal operation as well as during construction and periodic maintenance work. During maintenance, open hatches permit a high volume of air flow, combustible scaffolding adds to accumulated fuel, and cutting and welding activities act as the ignition sources.

Precautionary measures include designing and specifying equipment to minimize unwanted fuel accumulation, storing and handling fuels properly, using noncombustible construction, and providing adequate fire-suppression systems.

Coal can be safely stored when the following restrictions are adhered to:

- Coal should not be stored in contact with any source of external heat such as piping, flues, and boiler walls, or over steam mains.
- Fire service water-supply mains should not be run through coal storage areas.
- Coal storage should not be ventilated, as the draft or flue effect will aid a fire. Coal should

not be piled over manhole covers or covered pipe trenches, and vertical members such as timbers, columns, and large pipes should not pass through coal piles.

- Coal bins, silos, and bunkers should be constructed entirely of concrete or other noncombustible material. The structure should be roofed. A ventilation system should remove combustible gases given off by the coal.

- Coal fines should not be permitted to accumulate. Coal should be fed in at the top and withdrawn from the bottom of the structure.

- Coal storage structures should be sprinklered if they contain combustible material or serve other occupancies.

- Firefighting access should be provided in coal storage structures.

All pulverizing equipment between the inlet of the raw coal feeder and delivery of pulverized coal to the firebox should be designed to withstand an internal pressure of at least 50 psi (345 kPa; 3.45 bars) after allowing for expected corrosion and deterioration. Tensile stresses should not exceed approximately 60% of the yield strength or 25% of the ultimate strength. (Exceptions include portions of systems with adequate explosion venting, protected by explosion suppression systems, or operating under inert atmospheres.)

Electrical equipment in pulverizer areas should be listed for National Electric Code (NEC) Class II hazardous occupancy (see Sec. 94).

Easily-fractured coals (such as western coals) produce more dust (i.e., fines) than harder coals. When coal is transported on conveyors or dust is generated in coal crushing and pulverizing operations, *dust collection systems* must be used. *Dust suppression systems*, which spray a fine mist or foam on the coal where dust is generated, can be used. Enclosures to contain the dust, skirt modifications, and belt wipers should be used to minimize coal slippage on conveyors.

Automatic sprinkler systems and hose streams or hydrants should protect combustible conveyor belts. This is usually accomplished by having one supply line above the belt run and an unlimited number of sprinklers. Coal dust collectors should be protected with their own automatic spray systems, and they should incorporate explosion venting provisions (see Sec. 102).

Automatic sprinkler systems protecting conveyor belts should be hydraulically designed (see Sec. 41) to support 10 sprinklers and two small hose streams. Sprinkler spacing should be 100 ft^2 (9.3 m^2) per head. The

flowing pressure at the end sprinkler should not be less than 10 psi (70 kPa; 0.70 bars). (This will result in a combined sprinkler and hose stream demand of approximately 200 gpm (12.6 ℓ/s).) The water supply should be adequate for a minimum of 1 hour of operation.

Approved or listed $1\frac{1}{2}$-inch hose lines or hydrants should be provided at suitable intervals along the conveyor run.

The water sprinkler alarm (or other suitable fire detector) should shut down the conveyor drive when a fire is detected. Power to all contributing conveyors should also be shut off. Additionally, driving power to conveyors should be shut off when a 20% or greater belt slowdown or misalignment is detected.

Combustion gas mains should not be routed under or through the foundation (i.e., mat, slab, piers, and footings). Supply connections must be above grade and be sufficiently flexible to avoid rupture during an earthquake or differential settling of the foundation. Pressure regulators should be either located outside or the building or vented to the outside. Natural draft cross ventilation is necessary in all crawl and utility spaces containing gas service piping.

With modern methods of NO$_x$ control and the trend towards reduced amounts of excess air, there is an increased chance that unburned fuels will accumulate in the furnace during combustion upsets and flameouts. Therefore, the control system must be designed to trip the furnace whenever there is a loss of flame. Instrumentation to detect and measure carbon dioxide, combustion air, and flame is essential.

Flue gas desulfurization (FGD) systems (such as wet scrubbers) contain large amounts of combustible materials, including polypropylene and polyvinyl chloride (PVC) packing to protect the ducts from the highly corrosive vapors. When combustible linings are present, a specialized corrosion-resistant spray or other fire protection system is needed.

Fires in regenerative air heaters can be traced to overheated lubricating oil and rotor/air hood stoppage. Protective measures include monitoring inlet and outlet duct temperatures and rotor speed. A manually controlled water spray system (including standpipe), hose system, rotor observation ports, and hatches for fire fighting with hose streams should be included. Valves for drainage of firefighting water are also needed.

Filter fabric can be ignited by incompletely burned fuel particles. This hazard can be partially countered by the use of fabrics with high (i.e., in excess of 400°F (204°C)) temperature operating ranges. Other protective measures include automatic isolation valves and ducts to bypass the flue gas stream around the filter fabric, and a flue gas tempering water spray in the duct between the boiler and filter. As with air heaters, automatic spray systems and provisions for manual fire fighting should be included.

Fires in *electrostatic precipitators* (ESP) begin when products of incomplete combustion collect on plate surfaces and are ignited by electrical arcing. Such fires can be minimized by starting ESPs only after a stable furnace fire has been achieved. Fires can also start in electrical transformers. Temperature sensors should monitor the incoming and leaving duct temperatures. In addition to provisions for manual hose streams, fire barriers, spatial separation, and water spray systems should be incorporated.

Fluidized bed combustion (FBC) technology presents its own problems. Moisture can cause the calcium sulfite to become cement-like. Also, ash products tend to have greater carbon contents in some FBC designs than in other boiler designs. A greater buildup of combustible solids in the precipitators could result.

Stationary combustion engines and gas turbines can be protected by water spray, carbon dioxide, and Halon systems. Large individual turbines over 25,000 horsepower (18.63 MW) should have automatic sprinklers as a backup.

Water spray systems capable of supplying water to a design area of 2000 ft^2 (184 m^2) at a density of 0.2 gpm/ft^2 (8.1 ℓ/m$^2 \cdot$min) for 40 minutes can be used for isolated turbine installations.

Carbon dioxide systems intended to protect turbines in enclosed compartments should achieve a 34% concentration of carbon dioxide within 1 minute of actuation. A 30% concentration should be maintained throughout the deceleration period and until all metal surfaces have cooled below 400–500°F (204–260°C), the autoignition temperature of oil.

A 50% concentration of carbon dioxide should be achieved within 1 minute in accessory compartments where electrical and oil fires are possible. A 30% concentration should be maintained for the following 10 minutes.

Halon 1301 systems should maintain a 5% concentration during the deceleration period of the turbine. Halon should only be used where it can be established that fire temperature will not exceed the decomposition temperature—900°F (482°C) for Halon 1301.

The form of fire protection for oil-insulated transformers depends on the size and importance of the transformers. For example, a single transformer under 10,000 kVA probably can be protected by portable extinguishers. A single transformer over 10,000 kVA should have hydrant protection. Single transformers over 100,000 kVA should have a fixed automatic water spray system. Multiple transformers over 10,000 kVA should also be separated by 25 feet (8 m) clear space and/or noncombustible barriers between the units or be protected by a fixed water spray system.

Buildings and other structures near, or housing, oil-insulated transformers should be carefully considered.

14. Drainage

Four general methods are used to remove fire protection water from sites: grading, diking, trenching, and underground or enclosed drains. Various standards exist for specifying the storage capacity and design drainage rates for dikes and trenches. However, in general, the drainage system must be able to carry water away at a rate faster than it is applied. For example, a drainage trench must have the capacity to simultaneously remove water from fire hose discharges, the sprinkler system, rainfall, and all anticipated future developments. Open channel flow equations should be used for calculating the capacity of open trenches.

15. Record Keeping

Complete records help monitor the condition and readiness of a fire protection system. They are useful if the system is inspected by local, state, and federal officials. Such records also are necessary to limit liability in the event a lawsuit is brought against the building owner after a fire. Since there is no such thing as fireproof storage, in order for them to be available after a fire, records should not be kept in the building.

Three types of records should be kept. The first category includes the contractor's original shop drawings, specifications, manufacturers' operating and maintenance manuals, parts lists, and valve tag charts and diagrams. The second category of records includes all inspection and test records. The third category includes service and maintenance records.

Sprinkler Maintenance (NFPA 13A) should be consulted for more information on record keeping.

PART 2:
Detection and Warning Systems

16. Factors in Designing Fire Protection Systems

The first step in designing a fire protection system is to determine which codes and standards are applicable. This information can be readily determined by contacting the local building and fire officials.

Next, a clear understanding of the building's use and operation is required. In addition to knowing what the building will be used for, information about plumbing, HVAC, elevator, electrical, gas, and security systems is required.

Some equipment decisions are made on the basis of the construction budget. Other decisions are influenced by the knowledge and sophistication of the building's occupants.

Detector locations are influenced by many factors. Generally, the room designs, including HVAC, heating, and lighting designs, must be known before detector locations can be specified. Premature designs will result in inadequate protection, unreliable performance, and false alarms.

17. Approval of Equipment

All fire-protection products, particularly electrical equipment, installed in commercial and industrial buildings have to be approved by a nationally recognized testing laboratory (NRTL). NRTLs are accredited by OSHA.

Underwriters Laboratories (UL) and Factory Mutual Research Corporation (FM) have been performing such testing since the early 1900s and are the most widely recognized testing organizations. UL lists all varieties of electrical devices (including fire and explosion protection devices). FM is primarily involved with testing fire protection devices.

In 1988, OSHA published a regulation that permitted other companies, even those in other countries, to become NRTLs. Outside of UL and FM, only a few companies have become NRTLs. Some companies are limited to single testing sites; others are limited to testing certain types of products. UL and FM are accredited to test all product categories at all sites.

The approval granted to a piece of equipment by a NRTL can be an acceptance, test certification, listing, or labeling. The distinction between these forms of approval is confusing. An *accepted product* or *accepted installation* has been inspected and found by an NRTL to follow specific plans and code procedures.

A *certified product* has either been tested by an NRTL to nationally recognized standards, is safe to use in a specific way, has its manufacturing process regularly inspected by an NRTL, or has a certification mark (label or tag) affixed to it. A *listed product* is on a roster of products published by the NRTL, has been determined to meet national standards, and is safe for its intended purpose. A *labeled product* is identified with the label, mark, or symbol of the NRTL indicating that the product complies with recognized standards and the NRTL regularly inspects the manufacturing facilities where the equipment is made.

The standards against which the NRTL tests products must be approved by the Assistant Secretary of Labor. In general, ANSI and ASTM standards are used.

In addition to UL, FM, and a few others, some approvals have been granted by the Mine Safety and Health Administration (MSHA), formerly the Bureau of Mines. MSHA approvals have primarily dealt with methane atmospheres. However, equipment approved by MSHA for methane atmospheres cannot be used with any other hazardous material (even those in the same hazard groups) without additional approvals from UL or FM.

18. Primary Detectors

Primary detectors respond directly to a fire or its effects. Automatic detectors can detect heat, smoke, or flames.

Most detectors are wired into a parallel circuit or loop. However, some modern detectors are addressable. These include an analog data feature that permits continuous monitoring of sensitivity, detection of a failed or dirty sensor, and adjustment of sensitivity.

19. Manual Stations

Manual fire alarm (pull) stations, though not mandatory, should be considered for all commercial systems. Building configuration and the occupancy/hazard classification are two factors to be considered in locating pull stations. In general, they should be located in the natural exit path. The suggested distance between manual stations is less than 200 feet (60 m). When activated, identification of the station should be readily discernible down a corridor for at least 200 feet (60 m).

Manual pull stations can be either single-action switches (i.e., requiring a lever to be pulled) or double-action switches (i.e., requiring lifting a cover and pulling a lever). Once activated, most pull stations can be restored only by use of a special tool.

Due to vandalism, manual pull stations in some applications (e.g., schools) have been eliminated. A continuously monitored phone or other communication system is considered an acceptable substitute for manual stations in such cases.

20. Heat Detectors

There are three types of heat detectors: fixed-temperature, rate of rise, and rate-compensated.

Fixed-temperature heat detectors trip at a preestablished temperature no matter what the heat source is. The temperature is typically set at 135°F (57°C) for general spaces. In confined areas or rooms containing heat-generating equipment (e.g., boiler rooms, kitchens, etc.) and where normal heat buildup is expected to be severe or rapid, detectors operating in the 190–200°F (88–93°C) range are used.

Fixed-temperature heat detectors are constructed in one of three ways. The simplest detectors use a fusible alloy that melts at a predetermined temperature. Such detectors cannot be reset and must be bypassed or replaced when tripped. *Bimetallic detectors* use a bimetallic strip or disc. When the detector is heated, the difference in thermal expansion rates bends the strip (or snaps the disc from concave to convex) and closes the circuit. *Solid state detectors* use thermistors whose resistances change with temperature. A signal is generated when the resistance reaches a specific value.

Rate of rise detectors are more responsive than fixed-temperature detectors. They are tripped when either a fixed temperature, usually 135°F or 200°F (57°C or 93°C), is exceeded or when the temperature increases suddenly. The rate of rise portion of the detector operates when the temperature rises in excess of 15°F (8.3°C) per minute.

Rate-compensated heat detectors are considered to be the most responsive of all thermal detectors. They detect both slow- and fast-developing fires by anticipating the temperature increase and moving toward the alarm point as the temperature increases. As with other types of detectors, rate-compensated detectors are available in 135 and 200°F (57 and 93°C) models.

21. Smoke Detectors

Two main types of smoke detectors are in general use: photoelectric detectors and ionization detectors. Both are used for protection of large areas. However, each type has its advantages. The two types can be intermixed within a fire detection system to provide the best form of detection.

Photoelectric detectors are generally considered to be best for cold, smoldering (smoky) fires. The detectors operate on the principle that smoke particles interrupt and scatter a light beam, changing its normal intensity at the receptor. Photoelectric detectors respond best to products of combustion between approximately 0.3 and 10 microns (3×10^{-7} and 1×10^{-5} m) and in the proper concentration. Proper concentration is defined by UL as the ability to sense smoke in the obscuration range of 0.2–4.0% per foot.

A variation of the photoelectric detector is the *linear-beam smoke detector*. This is similar to the electric-eye door announcers still used in some stores. The light beam is projected across the area being protected toward a separately mounted receiver. Linear-beam smoke detectors are suitable for large areas (such as hotel atriums and churches) where products of combustion from small fires would be severely diluted.

Ionization detectors detect products of combustion by sensing the change of the air's conductivity in an ionized chamber. The ionization detector responds best to fast-burning fires where particle sizes range from 0.01–0.3 microns (1×10^{-8} to 3×10^{-7} m) in the proper concentration. Proper concentration is defined the same as for photoelectric detectors.

Less-used types of smoke detectors include resistance bridge detectors and air-sampling detectors.

Spacing of smoke detectors is covered in *Automatic Fire Detectors* (NFPA 72E). The area of protection should be reduced by 50% when destratification fans are used.

22. Flame Detectors

The two primary types of flame detectors are distinguished by the wavelengths they respond to. Both are line-of-sight devices requiring proper aiming and obstacle-free paths. Furthermore, the area being protected is limited by the inverse-square radiation law.

Infrared flame detectors respond to radiation with wavelengths in the 6500–8500 Å (6.5×10^{-7} to 8.5×10^{-7}m) range. Good detectors filter out solar interference and radiation from incandescent lamps and respond to radiation with wavelengths in the 4000–5500 Å (4×10^{-7} to 5.5×10^{-7} m) range. Detectors that have dual sensing circuits or that respond to flame flicker are preferable in order to minimize false alarms.

Ultraviolet flame detectors respond to radiation with wavelengths in the range of 1700–2900 Å (1.7×10^{-7} to 2.9×10^{-7} m). They can be used in areas that are illuminated by sunlight and incandescent lamps. Although these detectors are not susceptible to false alarms caused by solar interference, false alarms from other sources can be minimized by built-in time delays.

Two other types of flame detectors are flame flicker detectors and photoelectric flame detectors.

The circuitry for two or more types of flame detectors can be combined in a single unit. Then, two or more conditions must be present simultaneously for an alarm to be initiated.

23. Secondary Detectors

Secondary detectors (such as water-flow detector) do not detect fires or their immediate products. Rather, fires are detected from their effects on other systems.

Water-flow detectors are used in wet sprinkler systems and are triggered by the flow of water through the supply pipe. Such detectors are used on main sprinkler risers and elsewhere throughout a building. When properly zoned, the detector will indicate the floor or area in which the sprinkler discharge occurs. Retard (delay) mechanisms are employed to prevent false alarms from normally occurring water pressure fluctuations and surges.

Pressure switches signal a tripped valve by detecting a sudden change in pressure within a pipe. Such switches

are used in dry, preaction-action, and deluge systems, as well as in wet pipe sprinkler system alarm valves.

24. Miscellaneous Detectors

Gas detectors monitor the amount of flammable gases or vapors in an area. Strictly speaking, gas detectors do not detect fires—they detect the conditions necessary for a fire or explosion to occur.

Perimeters, pipelines, ductwork, and other building elements that extend over a distance can be protected by cable heat detectors (thermostatic cables). These detectors consist of two wires separated by a temperature-sensitive insulation. When a specified temperature is exceeded, the insulation melts and the wires short out (that section of the cable must then be replaced). The two wires can be oriented side by side (i.e., twin-lead) or can be concentric (i.e., coaxial).

Supervisory and *tamper switches* can be used to determine the readiness of a fire protection system without making a visual inspection. *Valve monitor switches* are used to detect a closure of water supply valves. Such valves are often closed for repair and maintenance and never reopened thereafter. As such, they constitute one of the most frequently forgotten parts of the system.

Water will not flow when it is frozen in the sprinkler pipes or when a pipe has burst. Therefore, low air temperatures (i.e., under 40°F (4°C)) are detected by low-temperature monitor switches.

Fire pumps, if used, should be supervised for all critical functions. This includes supervision of power to the pump, pump running, and other conditions (e.g., overheating and water supply) that would render the pump inoperable. The building fire alarm system should be provided with one or more zones to interface with fire pump signals.

25. Local Signal Devices

Local signaling devices are used to warn occupants of a fire threat in their building. The sound pressure level must be approximately 12 dB higher than the sound pressure level in the ambient environment.

A *bell* is the most commonly accepted type of local signaling device. However, bells should not be used in schools or other areas where bells are used for other signaling functions.

Chimes and single-stroke bells can be used in certain types of applications (e.g., quiet areas of hospitals and nursing homes). The strokes are coded (i.e., the frequency and timing of the strokes are distinctly different) for a fire alarm.

Sirens are effective signaling devices, but they are generally used only outside or in extremely noisy areas.

Buzzers are normally used to indicate nonfire conditions.

Speakers are used to give instructions during a fire. They are particularly useful in high-rise applications. The instructions can be provided by voice or by analog or digital recordings. Synthesized (computer-generated) voice instructions are used in rare instances.

Visual indicators (such as flashing lamps) can provide coded and noncoded signals. Such indicators may be mandatory in environments primarily designated for the handicapped. *Strobe lights* provide intense flashes of light and are effective in generating attention where low-visibility is a problem. Lights are almost always used in conjunction with some type of audible signaling device.

Remote annunciators duplicate the main control panel and have a light for every fire alarm zone. They are used at the main and secondary building entrances to assist the fire department in locating the fire zone.

26. Monitoring and Signaling Equipment

The purpose of the local monitoring/control panel is, first, to notify building occupants of the fire threat, and second, to notify a central station or the fire department. Fire and intrusion detectors should never be combined in one panel.

Depending on the building type, configuration, use, and hazard, there are many things that the panel could do upon receipt of a fire signal from a detector. Typically, the panel would perform some of the following operations:

- activate local signaling devices
- notify the fire department or central station
- signal the remote annunciator panels to indicate the location of the fire to arriving firefighters
- capture the elevators
- unlock all stairwell doors
- start elevator shaft and stairwell pressurization fans
- release all smoke door holders
- start exhaust fans and open HVAC exhaust dampers on the burning floor
- stop supply fans and close HVAC supply dampers on the burning floor

- stop exhaust fans and close HVAC exhaust dampers on adjacent (above and below) floors
- open HVAC supply dampers on adjacent (above and below) floors

27. Standby Panel Power

All modern fire alarm systems contain their own standby battery power for short-term outages. While standby generators can be used, battery power is the more common solution. Sufficient backup power must be provided to keep the detectors and control panel operating for at least 24 hours.

Unless required by local ordinance, a dedicated backup battery power supply is not necessary if the building has an approved emergency power system. The details of how any emergency generators feed power back to the building distribution system should be carefully considered to ensure emergency power in the event of a long-term outage. The *National Electric Code* (Articles 701 and 702), *Emergency and Standby Power Systems* (NFPA 110), and *Stored Emergency Systems* (NFPA 110A) are the standard references for designing private power-supply systems.

Unattended emergency generators should be protected by dedicated carbon dioxide systems.

28. Wiring and Zoning

While detectors can be connected with only two wires, supervision and addressability of the circuits require four wires to be used. Connections are almost always by hard wiring. Interconnections by radio, microwave, and fiberoptics are rarely used. Wire gauges of No. 18 AWG to No. 14 AWG are commonly used, although larger gauge wiring may be required by local codes or to limit resistance over long distances.

To assist fire fighters in rapidly locating the fire, detectors should be wired into zones. For example, each floor should be a different main zone; either the types of detectors (i.e., smoke, flame, or heat) or rooms should be subzones. The panel should have enough circuitry to be able to determine which zone is causing an alarm condition, to pass the zone information to the monitoring station, and to signal the annunciator panels at the building entrance.

29. Signaling

Communication between the local monitoring equipment and the fire department or central station can use a dedicated (leased) phone line, a standard voice phone line, or more rarely, radio transmission. Communication with on-site guard stations is usually by direct (hard-wired) connection.

Dedicated phone lines operating on the principle of reverse polarity transmit the fastest signals. However, dedicated lines are expensive.

Using standard phone lines requires a telephone dialer. Tape dialers have a bad reputation due to their high failure rate. Modern solid-state dialers have all but eliminated the objections to tape dialers. While the cost of a leased line is avoided, the notification takes longer because the panel must dial out and then wait for the call to be answered. Solid-state dialers are designed to take control of a phone line if the line is in use by another party.

30. Weatherability

None of the devices used in the fire detection and communication systems should be exposed to adverse conditions. Adverse conditions include excessive humidity, temperature extremes, corrosive atmospheres, and explosive atmospheres.

PART 3:
Fire Protection Methods

31. Fire and Explosion Control

Fires and explosions are controlled and extinguished by one of several techniques.

1. A fire may be extinguished by cooling the fuel. This is usually accomplished by applying water to a fire. However, cooling by ventilation air and inert gases can also be used.

2. Separating the oxidizer from the fuel will rapidly extinguish a fire. This method is used when blankets of foam or carbon dioxide (which is heavier than air) are applied. In water sprinkler systems, the steam produced from vaporized water separates the flame from atmospheric air.

3. The fuel supply can be diluted or removed. A blast of air will often extinguish fires involving combustible gases.

4. When sprinkler systems are used, the liquid fuels should be miscible with water. When liquid fuels are not miscible with water, extinguishment by water spray occurs by emulsification. The water spray should be applied over the entire burning surface layer. *Wet water* (i.e., water to which a compatible wetting agent has been added) should be considered.

5. Certain halogenated hydrocarbons and inorganic salts break the chain of chemical reactions that constitute combustion. This is known as *chemical extinguishment*.

6. Explosion suppression is similar to chemical extinguishment. A chemical that disrupts the explosion reaction is injected into the explosion within the first few milliseconds.

32. Portable Extinguishers

Water puts out most fires. It is the only option for truly large fires. However, it is not always the most effective extinguishing substance for small fires. It damages property and is dangerous when used around high-voltage electrical equipment. The threat and damage potential of using water near high voltages must be compared to the fire damage potential.

Table 4 classifies portable extinguishers into classes A, B, C, and D, depending on their effectiveness.[2] *Class A extinguishers* are for general purpose fires involving wood, cloth, paper, rubber, and many plastics. Class A extinguishers can also be used on electrical fires after the power has been shut off. *Class B extinguishers* are for flammable liquids, gases, and greases. *Class C extinguishers* are for electrical fires. Substances in class C extinguishers must be nonconducting. *Class D extinguishers* are for combustible metals such as magnesium, titanium, zirconium, sodium, and potassium.

Table 4
Portable Fire Extinguishers by Fire Class

extinguisher type	hazard class
foam	A
loaded stream	A, B
dry chemical	B, C
CO_2	B, C
vaporizing liquid	B, C
dry powder	D

Class A and B extinguishers are also numerically rated. The numerical ratings are developed from standardized tests. Table 5 can be used to evaluate the ratings. For

---

[2]Actually, the classification is for the extinguishing agent.

example, an extinguisher rated 4-A indicates that the extinguisher is approximately twice as effective in a class A fire as $2\frac{1}{2}$ gallons of water. Some extinguishing agents can be used on different fire types with varying effectiveness. Therefore, fire extinguishers are labeled with several ratings (e.g., 2-A:10-B:C).

Portable Fire Extinguishers (NFPA 10) should be consulted for additional information.

Table 5
Standard Ratings of Extinguishing Agents
Used in Portable Fire Extinguishers

description	rating
$2\frac{1}{2}$ gal water, under pressure	2-A
20 lb carbon dioxide	10-B:C
5 lb dry chemical (ammonium phosphate)	2-A:10-B:C
10 lb dry chemical (sodium bicarbonate)	60-B:C
10 lb dry chemical (potassium bicarbonate)	80-B:C
125 lb dry chemical (ammonium phosphate)	40-A:240-B:C
33 gal aqueous film-forming foam	20-A:160-B
5 lb Halon 1211	10-B:C
9 lb Halon 1211	1-A:10-B:C
1.5 lb Halon 1211/1301	1-B:C

Reprinted with permission from *NFPA Inspection Manual*, 6th ed., copyright © 1989, National Fire Protection Association, Quincy, MA 02269.

33. Dry Chemical Systems

Dry chemical systems are used primarily for small fires involving cooking exhaust systems, dip tanks, flammable liquids, and electrical equipment. Dry chemicals are frequently applied from portable hand extinguishers, propelled by pressurized gas discharges that place the chemicals in the combustion area.

The most common dry chemical agents are sodium bicarbonate, potassium bicarbonate (also known as PKP or Purple K), potassium chloride, urea-potassium bicarbonate, and monoammonium phosphate. Potassium bicarbonate is approximately twice as effective as the other chemicals on class B fires. Monoammonium phosphate is a multipurpose dry chemical and is useful on class A, B, and C fires.

When other options exist, dry chemicals should not be used on electronic equipment due to the difficulty in removing the chemicals from circuit boards and other devices.

34. Hose and Standpipe Systems

It is often desirable to direct water on a fire from within the building. This is accomplished with manually held hoses. While the fire department has the responsibility of fighting a fire, some early fire fighting can be performed by the building occupants. This is not recommended, however. Fire protection should never be based on the assumption that occupants will perform fire fighting services.

In high-rise buildings, outlets for hoses are supplied from standpipes.[3,4] For example, the UBC specifies (UBC Table 38-A) that multistory buildings over 150 feet (45 m) in height or buildings with four or more stories must have standpipes but not necessarily hoses.

Standpipes are pipes that run vertically through the building, providing water outlets for firefighting at each floor. They are often accessed from within the stairway of the building.

Standpipes can be wet or dry. *Wet standpipes* are always filled with water and are always ready for use. *Dry standpipes* are charged with water after a fire is detected. This can be done by pumps in the buildings or by the fire department using Siamese (i.e., wye or double) connections at the ground level. Both wet and dry systems can be refilled during use by the fire department from the ground level. Check valves in the system prevent water from being lost due to wrong outlets being opened and from excessive back pressure. (With the exception of check valves, however, Siamese connections should bypass all building valves and permit water to be injected directly into the standpipe system.)

A minimum pressure of 65 psig (450 kPa; 4.5 bars) is usually maintained at each hose outlet. Pressure-reducing valves can be used where necessary. For example, 100 psig (690 kPa; 6.9 bars) is the maximum pressure for $1\frac{1}{2}$-inch hose stations (NFPA 13, Sec. 4-4.1.7.23).

There are three classes of standpipes. Class I is a wet or dry standpipe system intended for use by fire department personnel, Class II is a wet or dry standpipe system intended for use by the building occupants, and Class III is a wet or dry system intended for use by both fire department personnel and occupants.

The distinction between occupants and fire personnel as intended users is based on the diameter of the connection provided. Outlets and hoses for occupant use are

---

[3]Standpipes can also supply sprinkler systems. However, the term *standpipe system* generally refers to a hose system.

[4]*Standpipe and Hose Systems* (NFPA 14) is the standard reference on standpipes.

smaller. The fire department connection is $2\frac{1}{2}$ inches (6.35 cm) in diameter, while the occupant connection is $1\frac{1}{2}$ inches (3.81 cm) in diameter. In the case of class III standpipes, the smaller $1\frac{1}{2}$-inch (3.81-cm) diameter hoses are used with the larger fittings with easily removable reducers.

In unsprinklered buildings, the UBC requires hoses to be supplied as parts of the standpipe outlets (UBC Table 38-A). Lengths must be sufficient to reach all parts of the floor or building covered. Each hose outlet is provided with a hose valve (i.e., an on/off shutoff valve). Sprinklered buildings generally do not require hoses.

Straight-stream hose nozzles are not typically installed for untrained users. However, nozzles capable of both spray (fog) and straight-stream operation are preferred where a fine spray is desired over large areas (such as where Class B hazardous liquids are being protected). Nozzles for hose lines are covered in Sec. 80.

Standpipes less than 100 feet (30 m) in height must be at least 4 inches (10.2 cm) in diameter. Those taller than 100 feet (30 m) but less than 275 feet (82.5 m) must be at least 6 inches (15.2 cm) in diameter.[5] Unless the system components have been pressure rated for greater pressures, standpipes may not be taller than 275 feet (82.5 m). For taller buildings, standpipes must be zoned to cover the entire building height.

Class II standpipes must flow at least 100 gpm (6.3 ℓ/s; 380 ℓ/min). Class I and III standpipes must flow at least 500 gpm (32 ℓ/s; 1 900 ℓ/min). Flow is at 65 psi (450 kPa; 4.5 bars). In fully sprinklered buildings, however, the total supply needs only to meet the larger of the standpipe or sprinkler needs.[6]

Water can be supplied to standpipes in two ways. Water storage-serving wet standpipes can be refilled through fire department pumps at ground level. Dry standpipes can be filled by the fire department through Siamese (wye) connections at ground level.

35. Foam Systems

Foams are masses of aqueous-based bubbles. When applied, a foam will blanket and exclude air from a fire. The bubbles float on all liquids. Foams also cool the burning material, remove heat when the water vaporizes, and reduce friction between rubbing components.

There are many types of foams, including protein foams, fluoroprotein foams, aqueous foams, film-forming foams (AFFF), alcohol foams, and high-expansion foams. High-expansion foams are particularly suited for indoor files.

Foam concentrates are mixed in small (i.e., 1–6% by volume) proportions with water. High-expansion foam is generated by flowing air through a screen that has been wetted with the foam concentrate.

The application rate depends on the type of foam, type of fire, and application device, but generally varies between 0.5 and 3 ft^3 per square foot of surface area (0.15–0.45 m^3/m^2).

Foam generation can be initiated automatically or manually on demand, as it is by airport emergency crews. Automatic systems are usually always activated by infrared or ultraviolet flame detectors. Such systems must be highly resistant to false alarms. Cross-zoning can be used to prevent unnecessary discharges.

36. Halon Systems

Halon is a general term that usually refers to several halogen gases containing bromine, fluorine, and carbon that are used as fire-extinguishing agents. Although many different Halon agents exist, most are either too expensive for common use, create toxic by-products upon decomposition in the fire, or have other deficiencies. Because Halons cause ozone depletion when released into the atmosphere, their production has been banned by international agreement in the United States and most other countries since 1994, though existing supplies may still be used.

Only two Halons are now in widespread use: Halon 1211 (bromochlorodifluoromethane, CF_2ClBr) and Halon 1301 (bromotrifluoromethane, CF_3Br). Both Halons are heavier than air, as is carbon dioxide. However, Halon 1301 is two and a half times more effective than carbon dioxide. Halon becomes ineffective at combustion temperatures above its decomposition temperature—900°F (482°C) for Halon 1301.

Halon systems are generally used in clean environments. Particularly suited are areas containing electronic equipment and other items that would be damaged by a water spray. In addition to computer rooms, Halon protection is provided for storage of magnetic tapes, microfilm, computer records, and underfloor spaces, as well as for research and medical laboratories. Halon protection is generally not extended to administrative applications such as word processing, personal computers, and personnel, payroll, and inventory records. Generally, the decision to use Halon protection is based on the value of the equipment or format of data being stored.

---

[5]These two limitations do not apply in completely sprinklered buildings when the standpipes have been sized by hydraulic calculations.

[6]Standpipe and automatic sprinkler systems should not be interconnected by sprinkler system piping (NFPA 13, App. A-4-4.1.7.24).

Halon 1301 is usually delivered to enclosed distributed hazards such as computer tape, data storage files, electronic equipment cabinets, relay switch vaults, and turbines by total flooding systems. However, it can also be used with local application systems.

Halon 1211 is used more frequently than 1301 in portable extinguishers due to its higher boiling point and other properties, although Halon 1301 is preferred in aircraft cabins and engines.

Halon 1301 is less toxic than Halon 1211. Halon 1301 becomes toxic to humans at concentrations of approximately 10% by volume in air, while Halon 1211 becomes toxic at approximately 4% by volume. It also requires approximately 10% less vapor than Halon 1211 for fire extinguishment. It can be used with total flooding applications in occupied spaces. Because of its toxicity, Halon 1211 should not normally be delivered by flooding in occupied spaces.

Halon systems, because of their high initial cost, are engineered for the particular rooms they are to protect. As with foam systems, cross-zoned detection systems and other techniques must be used to prevent false discharges.

Since July 1, 1992, releasing Halon gases has been permitted only in an actual fire. Even with a large reserve and provisions for recycling, use of Halon gases will eventually be curtailed. Alternative agents are under development.

37. Carbon Dioxide Systems

Carbon dioxide systems are used in industrial applications to suppress fires where, because of expense, hazard, or the size of equipment to be protected, water, extinguishing powders, and Halon would be unsuitable. These are typically type B and C fires, although carbon dioxide can be used on type-A fires. Carbon dioxide is particularly suited for fires involving flammable liquids and electric equipment. Also, it is useful when large volumes are to be protected. However, since concentrations above 30–40% by volume are required and concentrations above 9% by volume constitute a suffocation hazard, the use of carbon dioxide systems must be carefully considered.

Carbon dioxide is considered a temporary extinguishing agent. After the gas has dissipated, a fire may reignite. Therefore, a second type of agent is needed to permanently extinguish deep-seating glowing embers.

Carbon dioxide can be applied through total area flooding, fixed-spot equipment, hand hose line, and standpipes with mobile supply.

38. Fixed Water Sprinkler Systems

Fire departments are seldom able to direct ground-mounted water streams to heights greater than approximately six stories (approximately 75 feet (23 m)). Therefore, buildings above this height must provide their own fire protection systems. Fixed water sprinkler systems satisfy this requirement. For example, the UBC requires sprinklers in buildings over 75 feet (23 m) high (UBC Sec. 3802). Other restrictions (see Sec. 34) are placed on buildings higher than 150 feet (45 m).

Water sprinkler systems consist of a horizontal network of pipes located near the ceiling. By definition, each sprinkler system is identified by its own single system riser control valve (NFPA 13, App. A-1-4.2). Sprinkler heads for ordinary use are constructed so that a temperature in the range of 135–170°F (57–77°C) will cause them to open (see Sec. 59).

In tall buildings, sprinkler systems are supplied by either water tanks near the top of the building or, more frequently, pumps (with suitable emergency power) drawing from sources at the ground level. Tanks should be designed to supply a specific number of sprinklers for a specified time. The tanks should also be capable of being refilled from the exterior by fire department pump trucks. This is accomplished through exterior Siamese (wye) connections. (See Sec. 55.)

Sprinkler systems can be either of the wet pipe or dry pipe variety. The pipes of *wet pipe sprinkler systems* are constantly filled with water. Any triggered sprinkler head will immediately discharge. *Dry pipe sprinkler systems* are typically used in unheated buildings and attic spaces where the water in the pipes might otherwise freeze. Dry pipe systems are charged with air or nitrogen. Dry valves with 20 psig (140 kPa; 1.4 bars) air pressure can hold back 120 psig (830 kPa; 8.3 bars) of water. When a spray nozzle is opened, the air pressure is reduced, the valve opens, and water flows from the mains into the distribution system.

In order to prevent freezing, wet sprinkler systems must be kept above 40°F (4°C) (NFPA 13, Sec. 4-5.4.1). When a danger of freezing is present and no other alternatives (e.g., providing pipe heat or using a dry pipe system) exist, wet sprinkler systems can be charged with an inexpensive antifreeze solution such as glycerine or propylene glycol. Systems containing some antifreeze solutions should not be directly connected to the municipal water system without approved backflow preventers (NFPA 13, Sec. 3-5.2.1).[7]

Because of cost, complexity, and environmental concerns, systems with antifreeze are seldom used when large areas are to be protected (e.g., when there are more than approximately 20 sprinklers or when more than 40 gallons (151 ℓ) of antifreeze solution are involved). Dry pipe systems should be used when large areas are to be protected.

39. Preaction Sprinkler Systems

Preaction dry pipe systems are actuated by either open sprinklers, supplemental sensors (i.e., flame, smoke, or heat sensors), or both (NFPA-13, Sec. 3-3.2.1).[8] When a fire is detected, water is released into the pipe system, but there is still no discharge unless a sprinkler head has actually been opened. Preaction systems have an advantage of not tripping when accidental pipe or sprinkler damage occurs. However, their primary advantage is that the same sensors that open the water valve can send a signal to the control panel before water is discharged. This may permit the fire to be put out without incurring additional water damage.

A very important consideration in choosing to use a preaction system is the reliability. Because preaction systems are more complex, there are more system elements that can fail and prevent proper operation. Preaction systems are considered to be much less reliable than standard water spray sprinkler systems. Careful engineering design is required to ensure adequate reliability.

40. Deluge Systems

Deluge water sprinkler systems are a special category of water spray systems. They are used to protect areas (e.g., airplane hangars and warehouses containing propellants and explosives) where the risk of a fast-acting, widespread fire (i.e., flash fire) is significant and regular sprinklers would not act quickly enough over a large enough area. The deluge valve can be actuated manually, but it is usually triggered by thermal or flame detectors. Since there are no fusible elements in the sprinklers, the entire protected area receives spray simultaneously from all sprinklers.

High-speed preprimed deluge systems, which become fully operational within 0.5 seconds, should be used wherever exposed powder, explosives, or propellants are processed or stored.

Deluge systems may discharge 5000 gpm (320 ℓ/s; 19 000 ℓ/min) or more. Such systems seldom can be supplied by the municipal water system. On-site water supply tanks or other sources are required.

---

[7]The local building, health, and fire officials may permit a system containing glycerine or propylene glycol to be connected to the municipal water system through approved double-check valves and backflow preventers. However, diethylene glycol, ethylene glycol, and calcium chloride systems cannot be so connected (NFPA 13, Sec. 3-5.2.1). Also, CPVC pipe is damaged by glycol-based compounds and should be protected by only glycerine (NFPA 13, App. A-3-5.2).

[8]Systems requiring both sensors and sprinklers to be triggered are known as *double interlock preaction systems*.

PART 4:
Design of Water Sprinkler Systems

41. Schedule- vs. Hydraulically-Designed Sprinkler Systems

Pipes used in sprinkler systems can be sized in one of two ways. Before calculators and computers made sizing calculations simpler, the traditional manner was to use a *pipe schedule*—a table (such as Table 11) dictating the maximum number of fire sprinklers that can be served by any size of pipe. Different occupancy hazards call for different schedules. This method of design has been in use for more than a century.

With a few exceptions, schedule design is permitted only on new construction when less than 5000 ft^2 (465 m^2) is to be protected and when existing schedule-designed systems are expanded (NFPA 13, Secs. 5-2.2.1 and 6-5).[9] Furthermore, schedule designs are permitted only with sprinklers that have the standard 0.5-inch (1.3-cm) diameter orifices and with listed ferrous and copper piping materials (NFPA 13, Sec. 6-5.1).

Most contemporary sprinkler system designs (and all designs for deluge and water spray systems) are based on hydraulic calculations. This method of design is covered in Secs. 75–78. The total water supply requirements for hydraulically-designed systems are lower than for schedule-designed systems. Also, pipe sizes can be reduced (down the run) in hydraulically-designed systems. For these two reasons, hydraulically-designed systems are generally more economical to install.

42. Sprinkler vs. Water Spray Systems

Water spray sprinkler systems can be thought of as spot protection and are similar to deluge systems in concept and operation. The discharge can be triggered manually or by sensor. The nature of the protection (i.e., the volume, type of spray, and area covered) depends on the type of hazard being protected against.[10]

Water spray systems are not substitutes for sprinkler systems. Sprinkler systems protect the building and structure. Spray systems protect specific hazards such as tanks of flammable liquids and gases, electrical transformers, and rotating electrical machinery.

43. Conventional vs. Residential Sprinklers

Conventional (i.e., commercial and industrial) sprinkler systems are usually unsightly, expensive, and require too much ceiling space to be used in residences. Also, conventional sprinklers achieve their best coverage with ceiling heights higher than the normal residence. Finally, conventional sprinklers react too slowly for residential fires, where heat and toxic gas levels would reach lethal levels before the sprinklers opened. Therefore, conventional sprinkler systems are seldom installed in residences.

Since the late 1970s, new developments have made residential sprinkler systems more attractive (i.e., less expensive).[11] These developments include low-cost piping and quick-response sprinklers with modified spray patterns. Unless they are specifically intended for dry operation, residential sprinklers should only be used in wet systems (NFPA 13, Sec. 4-3.6.2).

---

[9]The 5000 ft^2 (465 m^2) limitation does not apply when sufficient flow (as defined in Table 7) is available with a residual pressure of 50 psi (345 kPa; 3.45 bars) at the highest sprinkler (NFPA 13, Sec. 5-2.2.1).

[10]Design of water spray systems is covered in *Water Spray Fixed Systems* (NFPA 15).

[11]Design of residential sprinkler systems is covered in *Standard for the Installation of Sprinkler Systems in One- and Two-Family Dwellings and Mobile Homes* (NFPA 13D) and *Standard for the Installation of Sprinkler Systems in Residential Occupancies up to and Including Four Stories in Height* (NFPA 13R).

44. Tree vs. Loop Sprinkler Systems

Pipe networks can be designed as tree, gridded, or loop systems. With common *tree systems*, the first sprinkler to operate discharges at a greater-than-design rate due to the nature of the flow's declining pressure. *Loop and gridded systems* cannot eliminate the flow-related friction loss, but by providing multiple (parallel) paths to an open sprinkler, the friction loss can be reduced or minimized. Therefore, sprinkler pipe sizes can be smaller (or sprinkler spacing can be larger) with loop systems than with tree systems.

Although loop and gridded designs enjoy a slight economic benefit from needing fewer sprinklers or smaller piping, they require more piping and more complex design calculations. The cost of materials may be slightly greater for loop systems than for tree systems. With modern computer design methods, design costs are probably not an important factor.

A *circulating closed-loop system*, which is a permitted application, is a wet pipe-looped sprinkler system in which the pipes have a second purpose—usually to carry heating or cooling water. Such systems are also known as *automatic sprinkler systems with nonfire protection connections*. Water is only circulated through the system; it is not used or removed from the system (NFPA 13, Sec 3-6).

Gridded systems are commonly used when the protected area is large and roughly rectangular.

45. Responsibility for Design

Designing sprinkler systems should be left to knowledgeable specialists. Although any engineer will be familiar with the hydraulic design concepts necessary to design a sprinkler system, many different codes and regulations need to be followed. Designs will vary depending on the facility to be protected. For example, commercial buildings, aircraft hangers, and rubber tire storage facilities are treated differently. For most residential and commercial buildings, *Standard for the Installation of Sprinkler Systems* (NFPA 13) should be consulted.

The current state of the art is that the sprinkler system is a design-built feature of a building. Sprinkler system layouts are not normally included in the contract drawings. The architect and design engineer do not specify all of the details of the system. While certain elements (e.g., sources of water and power) must be provided in the design by the architect or engineer, others are not.

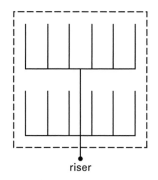

(a) tree system (side central feed shown)

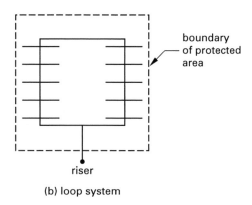

(b) loop system

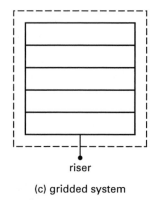

(c) gridded system

Figure 1   Tree, Loop, and Gridded Pipe Systems

In buildings with simple configurations and low hazards, details such as the locations of sprinkler pipe and sprinklers, types of sprinklers and hangers, and other methods of installation are left to a sprinkler contractor (i.e., the sprinkler installer). In more complex buildings or where hazards are high, a fire protection engineer will contribute to the design. Shop drawings are submitted by the sprinkler system designer for approval by the contract administrator and local authorities.

In many situations (and when permitted by law), the local building and fire officials will have the ultimate authority in determining if the system meets the applicable codes.

46. Sprinkler System Diagrams

Descriptions of sprinkler systems make use of the following component terms:

- *branch line*: a pipe containing sprinklers crossing a cross-main

- *cross-main*: a pipe directly supplying cross lines containing sprinklers

- *feed main*: a supply riser or a cross-main

- *riser*: a pipe, approximately vertical, usually extending one full story

- *system riser*: an above-ground pipe, approximately vertical, bringing water from the supply source

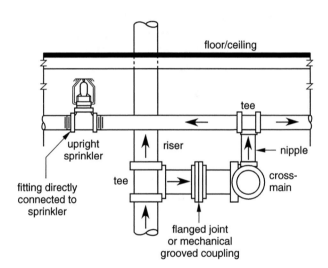

Figure 2 Typical Sprinkler System Components

Diagrams of sprinkler systems, such as Fig. 3, make use of the abbreviations listed in Table 6. Symbols for the various components are shown in App. G.

47. Sprinkler System Materials

All materials, pipes, sprinklers, and electrical devices used in fire protection systems should be new (NFPA 13, Sec. 2-2.1.). Where applicable, they should also be approved and/or listed (NFPA 13, Sec. 2-1.1).

48. Types of Allowed Pipe

Piping used in sprinkler installations must be ferrous pipe (welded or seamless), electric resistance-welded steel pipe (schedule-40 or thin-wall schedule-10), wrought steel pipe, or in some cases, listed copper tube and water tubing (types K, L, and M only), and when

suitable, nonmetallic pipe such as polybutylene and chlorinated polyvinyl chloride (CPVC) (NFPA 13, Sec. 2-3.1).[12]

Table 6
Standard Abbreviations for Sprinkler Diagrams

ALV	alarm valve
BV	butterfly (wafer) check valve
Cr	cross
CV	swing check valve
Del V	deluge valve
DPV	dry-pipe valve
EP	90° elbow
EE	45° elbow
GV	gate valve
Lt.E	long-turn elbow
OS&Y	outside stem (screw) and yoke manual control valve
PIV	post indicator valve (control valve)
St	strainer
TP	tee, flow turned 90°
WCV	butterfly (wafer) check valve

Reprinted with permission from NFPA 13, *Standard for the Installation of Sprinkler Systems*, copyright © 1991, National Fire Protection Association, Quincy, MA 02269. This reprinted material is not the complete and official position of the National Fire Protection Association on the referenced subject, which is represented only by the standard in its entirety.

The minimum working pressure (i.e., pressure rating) is 175 psi (1.21 MPa; 12.1 bars) for all pipe materials (NFPA 13, Sec. 2-1.2). The words *pipe* and *tube* are considered to be synonymous. No ferrous pipe should be less than 1-inch (2.54-cm) nominal diameter; copper tubing should be at least $\frac{3}{4}$-inches (1.9-cm) in nominal diameter (NFPA 13, Sec. 6-4.1). For steel pipe, schedules 10, 30, and 40 are commonly used.

For underground use, water pipes are normally class 150 cast iron with a working pressure of 150 psi (1.035 MPa). However, it is not a good idea to operate at that pressure.

49. Pipe Hangers and Braces

Hangers used in sprinkler systems must be capable of supporting the pipes and their water contents. Strap hangers are frequently used. Each hanger (including

---

[12]Plastic pipe's suitability for use in fire depends on two factors: cooling and protection. Water discharged from sprinklers provides the primary cooling effect. Since a fire may be in existence for a short period before sprinklers are triggered, plastic pipe should be protected by some form of fire-resistant membrane. Lay-in drop ceiling panels constitute such a membrane, but the protection might be lost in an earthquake or if ceiling panels are missing. Exposed plastic pipe is permitted with fast-response sprinklers in light hazard occupancies and when listed by the pipe manufacturer for use in specific ordinary hazard occupancies.

its support point and the structural system) must be capable of supporting 250 pounds (113 kg) plus five times the weight of the water-filled pipe tributary to the support (NFPA 13, Sec. 2-6.1). All hangers must be constructed of ferrous materials (NFPA 13, Sec. 2-6.1.2).

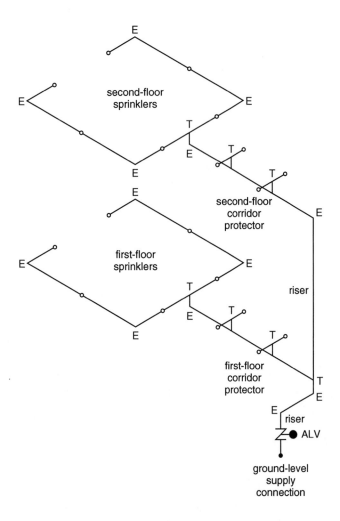

Figure 3   Typical Fire Sprinkler System

In seismically active areas, special protection against earthquakes is required. This is accomplished by designing bracing and flexibility into the systems. Flexibility is achieved by using listed flexible couplings and grooved end pipes and by allowing clearances. However, excessive flexibility is not permitted (NFPA 13, Sec 4-5.4.3). Piping 3 inches (7.6 cm) or smaller is considered flexible enough to not need flexible couplings (NFPA 13, App. A-4-5.4.3.2).

Bracing is required for mains and the ends of branch lines (NFPA 13, Sec. 4-5.4.3.5.8). Hangers must be sized to resist earthquake-induced accelerations of one-half gravity in both vertical directions. This means that long-column buckling failure must be checked as well.

To prevent buckling failures, braces cannot have slenderness ratios greater than 200 (NFPA 13, Sec. 4-5.4.3.5.1).

CPVC and polybutylene pipe are particularly susceptible to strain distortion from water reaction. Special support requirements apply to pipes of these materials (NFPA 13, App. A-4-5.2.3.3).

Other standards for placement and maximum distance between hangers must be met (NFPA 13, Secs. 4-5.2.2 and 4-5.2.3).

50. Valves

Control valves are used to connect two water supplies and to connect water supplies to sprinkler systems. Because of their importance, control valves in sprinkler systems must indicate "open" or "closed" in some manner (NFPA 13, Secs. 2-7.1.1 and 4-5.1.1.1). They should be either post-indicating (PI) or outside-stem-and-yoke (OS&Y) types. No control valves are permitted in connections intended for use by fire departments (NFPA 13, Sec. 4-5.1.1.2). All new valves must be right-hand valves.

All water-supply valves must either be electronically supervised (i.e., monitored), locked open, or isolated in a secured (but regularly inspected) location (NFPA 13, Sec. 4-5.1.1.3).

In order to prevent damage from *water hammer*, valves must not close in less than 5 seconds when operating at maximum speed from a fully opened position (NFPA 13, Sec. 2-7.1.1).

Drain and check valves must be specified to handle the same pressure as the piping: 175 psi (1.21 MPa) (NFPA 13, Sec. 2-7.1.2).

Alarm valves are designed to produce a signal when a flow equivalent to a single sprinkler head for 5 minutes is achieved (NFPA 13, Sec. 2-9.1). They are required on sprinkler systems with more than 20 sprinklers (NFPA 13, Sec. 4-6.1.1.1).

Check valves are needed when there is more than one source of water supply (NFPA 13, Sec. 4-5.1.1.4). Double-check valves are often used to achieve critical isolation. Check valves are designed to be installed either horizontally or vertically. Vertical check valves should not be installed horizontally and vice versa (NFPA 13, Sec. 4-5.1.1.5).

All valves should be identified by tags. The identification should indicate the purpose of the valve.

All valves should be accessible by authorized personnel during an emergency. This may require providing permanent ladders, stair treads clamped on risers, and chain-operated control wheels.

51. Pressure-Reducing and Pressure-Regulating Devices[13]

Pressure-regulating valves (PRVs) and devices reduce static pressure under both flow and nonflow conditions. Many sprinkler components are rated at 175 psi (1.21 MPa; 12.1 bars). Therefore, pressure-regulating devices set for 165 psi (1.14 MPa; 11.4 bars) are used where it is expected that the water pressure will exceed that limit (NFPA 13, Sec. 4-5.1.2.1). Pressure gauges are installed on both sides of the device (NFPA 13, Sec. 4-5.1.2.2).

Use of PRVs is restricted by *Private Fire Service Mains* (NFPA 24), which should be consulted. Where essential, PRVs should be installed on individual services rather than on main piping. When used with sprinkler systems and other systems intended for fire protection, PRVs should have control valves on each side, a bypass around the valve, and a relief valve on the discharge (low-pressure) side. The relief orifice should be at least $\frac{1}{2}$-inch in diameter, and the valve should be set for no higher than 175 psi (1.21 MPa; 12.1 bars).

A *pressure-reducing device* reduces pressure under flow conditions only. For pressures above approximately 100 psi (690 kPa; 6.9 bars), disks with restricting orifices are used. Hose systems are usually rated at 100 psi (690 kPa; 6.9 bars). Pressure reducing devices are required for even higher supply pressures (i.e., above 175 psi).

52. Sprinkler Supply Parameters vs. Occupancy Type

In a fire, several systems may be in operation simultaneously. For example, sprinklers, internal hoses, and fire department connections may all be used. The water supply flow rate and pressure must remain within the acceptable range for their durations of use. The required supply is a combination of sprinkler demand and hose allowances.

The sprinkler supply requirement can be determined either from area/density curves (Sec. 63) or from the room design method (NFPA 13, Sec. 5-3.3.1.2). For most installations with multiple rooms, compartments, and areas, it is not necessary to base the design flow rate on all sprinklers in all areas operating simultaneously. The basis of the room design method is that when certain conditions are met, the water-supply requirements

are based on the room and communicating space, if any, that generate the greatest hydraulic demand (NFPA 13, Secs. 5-2.3.3.1, 6-4.4.1(b)).[14] Corridors are considered to be rooms (NFPA 13, App. A-5-2.3.3.1).

Unless an optimizing computerized calculation is used, at least three sets of manual calculations covering different trial areas are required to prove an area protected by a gridded sprinkler system is hydraulically most-demanding (NFPA 13, Sec. 6-4.4.2).

The building occupancy hazard type (Sec. 3) determines the maximum protected area per sprinkler, required pressure, required flow rate, and duration of flow. A distinction is made between schedule- and hydraulically-designed systems (see Sec. 75). Typical values for schedule systems are specified in Table 7.

The maximum floor area on any single floor that can be protected by a single sprinkler system riser is 52,000 ft^2 (4831 m^2) for light and ordinary hazards. The limit for extra hazard occupancies is 25,000 ft^2 (2323 m^2) when the design is by pipe schedule and 40,000 ft^2 (3716 m^2) when the design is hydraulically calculated (NFPA 13, Sec. 4-2.1).

The supply requirement for hydraulically-designed systems is calculated from the sprinkler demand and an allowance for inside and outside hose streams. The supply must be available for the minimum duration shown in Table 7 (NFPA 13, Sec. 5-2.3.1.1). The sprinkler demand can be found from either area/density curves (see Fig. 9) or the room design method (NFPA 13, Sec. 5-2.3.1.2). Allowances for hose streams are covered in Sec. 79.

The supply curve and system loss curve can be plotted on the same graph to determine if the supply can provide the necessary pressure at the required flow. (It is common to use a semilogarithmic graph paper and plot the discharge quantity, Q, raised to the 1.85 power.) Starting at the static elevation head (i.e., the height of the highest sprinkler above the supply), the sprinkler demand curve is drawn based on the friction drop and required flow rate. The system curve should not intersect the supply curve. The horizontal distance from the operating point to the supply curve represents the volume available for hose streams. This must exceed the quantities shown in Table 16.

---

[13]The distinction between pressure regulating and pressure reducing valves is not consistently made. NFPA 13, for example, makes reference only to pressure-reducing valves.

[14]In order to use the room design method, all rooms must have fire-resistant walls with time ratings equal to or longer than the water supply durations indicated in Table 16 (NFPA 13, Sec. 5-2.3.3.1).

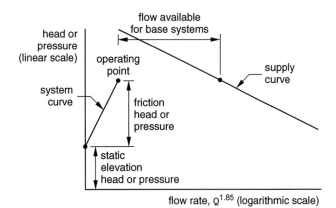

Figure 4   System and Supply Curves

With a properly designed wet system, the fire department demand is not added to the supply requirement (NFPA 13, Sec. 5-2.3.1.3(c)). Allowances for inside and outside hose streams are not necessary when gravity or pressure tanks or pumps supply only sprinklers (NFPA 13, Sec. 5-2.3.1.1, Exceptions 2 and 3; Sec. 5-2.3.1.3(h)).

When one set of sprinklers protects two layers (e.g., when a sprinkler system protects both the floor below and the plenum area above a dropped ceiling), the larger demand shall be used to size mains and supply lines (NFPA 13, Sec. 6-4.4.3(b)).

Table 7
Supply Parameters vs. Occupancy Type
(scheduled sprinkler systems, exclusive of hose streams)

occupancy classification	residual pressure required (psi)	acceptable flow at base of riser (gpm)	required duration (min)
light	15	500–700	30–60
ordinary	20	850–1500	60–90

Multiply psi by 6.895 to obtain kPa.
Multiply psi by 0.0689 to obtain bars.
Multiply gpm by 0.0631 to obtain ℓ/s.
Multiply gpm by 3.785 to obtain ℓ/min.

Reprinted with permission from NFPA 13, *Standard for the Installation of Sprinkler Systems*, copyright © 1991, National Fire Protection Association, Quincy, MA 02269. This reprinted material is not the complete and official position of the National Fire Protection Association on the referenced subject, which is represented only by the standard in its entirety.

In Table 7, the stated pressure is required at the highest sprinkler (sometimes referred to as the *pressure at the roof*) exclusive of the elevation and friction pressure losses to that point. The pressure at the base of the riser supplying sprinklers above will include the elevation head.

The lower flow rate given in Table 7 is the minimum flow rate, including fire hose streams, and is permitted only when (1) the building is of noncombustible construction, or (2) the potential for fire is limited to 3000 ft^2 (279 m^2) for light hazard occupancies and 4000 ft^2 (372 m^2) for ordinary hazard occupancies by building size or compartmentalization. Even when the lower rate is permitted, the higher flow rate should be chosen when a conservative approach seems warranted (NFPA 13, Sec. 5-2.2.4).

The lower time duration given in Table 7 is ordinarily acceptable when a water flow alarm reporting to a remote or central station is provided (NFPA 13, Sec. 5-2.3.1.3 (g)). The higher duration will be acceptable in all installations.

Generally, the static supply pressure at the system inlet should be at least 50 psi (345 kPa; 3.45 bars) or greater. The actual value will depend on the expected friction losses.

53. Sources of Water for Firefighting

Sources of water for firefighting and resupply should be reliable (e.g., municipal water supplies, gravity water tanks, and fire pumps with adequate supplies). Natural sources, such as surface lakes, rivers, and reservoirs are often considered to be unlimited, although they may be limited by freezing weather, drought, and evaporation. Ground water sources (such as wells and springs) can be similarly limited. *Drawdown* is the decrease in the elevation of the water level in the well, tank, or sump during continuous pumping operation.

A *cycle system* is one that is able to collect, strain and clean, and reuse applied water. Collector trenches or interceptor systems are required. Cycle systems may be permitted where the water supply is extremely limited.

Cross connection with multiple sources is desirable. However, if a local or industrial water source is connected to a municipal source, approved backflow preventers and check valves are required to prevent contamination of the municipal water system.

Static pressures in most municipal systems are in the 65–75 psi (450–520 kPa; 4.5–5.2 bar) range. A minimum pressure of 20 psi (140 kPa; 1.4 bars) should be available at flowing hydrants and at city mains when a fire pump is running at 150% of capacity.

The duration that water flow must be available is an engineering judgment, with input from the local building and fire officials.

54. Flow Measurement in Sprinkler Systems

Water taken by a fire protection system from a municipal system may or may not be metered. Bypassing the metering device ensures that the highest possible water pressure will be available at the sprinkler heads. However, if the municipal water supply pressure is high enough, this reasoning may not be sufficient to avoid metering sprinkler water.

The connection to the water supply can be direct, or a special bypass meter valve can be used. Low quantities (i.e., normal use) of water flow through the meter. Large quantities automatically flow straight through the valve.

55. Water Tanks

Water tanks are classified as gravity tanks, ground (suction) tanks, and pressure tanks. Actual required volumes are calculated from the required (or calculated) demand rate and the required duration.

Gravity tanks can be made of steel, concrete, or (rarely, nowadays) wood.[15] The capacity of steel tanks is usually in the 5000–500,000 gallon (20–2000 m^3) range. Wood tanks are usually limited to approximately 100,000 gallons (400 m^3).

Gravity tanks are located on building rooftops or mounted on towers. Tanks should be at least 35 feet (10.5 m) above the highest sprinkler served. This will ensure a 15 psi (105 kPa; 1.05 bars) minimum static pressure at the highest sprinkler.[16] (NFPA 13 Sec. 6-4.4.8 requires a minimum pressure of 7 psig (48 kPa; 0.48 bars) at each sprinkler.) When water tanks serve both sprinklers and hose streams, the water tank outlet should be even higher, approximately 75–150 feet (23–45 m).

Ground tanks are set on the ground and generally feed supply pumps. For this reason, they are often called *suction tanks*. Ground tanks can be constructed of steel, concrete, or wood, although rubberized-fabric tanks (i.e., bladders) are increasingly used. Rubberized-fabric tanks are available in 20,000–50,000-gallon sizes (in increments of 10,000 gallons) and in 100,000-gallon increments up to 1 million gallons.

Pressure (hydropneumatic) steel water tanks are expensive and generally are unable to deliver adequate volume for large areas.[17] However, pressure water tanks are

---

[15]NFPA 22, *Standard for Water Tanks for Private Fire Protection* is the accepted reference for gravity water tanks.

[16]Actually, the friction loss must also be considered.

[17]NFPA 22, *Standard for Water Tanks for Private Fire Protection* is the accepted reference for pressure tanks.

acceptable supplies for sprinklers and hand hoses, particularly for spot protection of limited duration (NFPA 13, Secs. 7-2.3.1.1 and 7-2.3.1.3).

The volume of a pressure tank must be adequate to fill dry pipe and preaction systems in addition to supplying the sprinklers for the minimum duration (NFPA 13, Sec. 7-2.3.2). Typical commercial capacities are in the 2000–10,000 gallon (7.5–40 m^3) range, although other volumes can be manufactured to specification.

When the tank bottom is above the highest sprinkler, the required air pressure in a pressure tank is calculated from Eq. 1 (NFPA 13, App. A-7-2.3.3). p_{outlet} is the tank's outlet pressure, in psi, as determined from the sprinkler system's required inlet pressure. In the absence of other information, the outlet pressure is taken as 15 psig. The fraction of air in the tank, A, is normally one-third (i.e., the tank is normally kept approximately two-thirds full of water). However, the fraction of air normally in the tank can be increased to maintain a higher discharge pressure.

$$p_{air} = \frac{p_{outlet} + 15}{A} - 15 \qquad \text{[U.S.]} \quad 1(a)$$

$$p_{air} = \frac{p_{outlet} + 103}{A} - 103 \qquad \text{[SI]} \quad 1(b)$$

$$p_{air} = \frac{p_{outlet} + 1.03}{A} - 1.03 \qquad \text{[}p \text{ in bars]} \quad 1(c)$$

The air pressure in the tank must be at least 75 psi (520 kPa; 5.2 bars), which is sufficient to maintain a minimum water pressure of 15 psi (105 kPa; 1.05 bars) at the highest sprinkler when flowing (NFPA 13, Sec. 7-2.3.3). This corresponds to $A = \frac{1}{3}$ for a tank two-thirds full of water.

If the bottom of the tank is below the highest sprinkler, the tank pressure must be at least 75 psig (520 kPa; 5.2 bars) plus the static pressure head corresponding to the difference in elevations between the highest sprinkler and the tank bottom divided by A (NFPA 13, App. A-7-2.3.3). When the tank is one-third full of air, this requirement translates into a pressure increase of three times the static pressure head corresponding to the difference in elevations between the highest sprinkler and the tank bottom (NFPA 13, Sec. 7-2.3.3).

Pressure tanks must have the means to automatically maintain the required pressure. As with a pump, when a pressure tank is the only water supply, it must be supervised (NFPA 13, Sec. 7-2.3.1.2).

56. Fire Hydrants

Street *fire hydrants* are usually installed by municipalities, but in large multiacre plants and developments, the developer will be responsible for installation. Hydrants are not normally designed. Rather, they are chosen by fire protection engineers from standard available designs on the basis of their required operating characteristics. All fire hydrants selected should be listed or approved.

Since modern fire truck pumps are capable of flowing 1000 gpm (63 ℓ/s; 3800 ℓ/min) or more, hydrants with lower-rated capacities (e.g., 500 gpm or 750 gpm) may limit firefighting efforts. Suggested minimum performance criteria of hydrants are:

- one large (4-, $4\frac{1}{2}$-, or 5-inch) pumper (steamer) connection for supply to pumpers

- two hose connections, typically $2\frac{1}{2}$ inches

- 1000-gpm flow rate when other hydrants are open

- 20-psi residual pressure

- wet barrel (where there is no danger of freezing)

Hydrants are assumed to deliver water at a residual pressure of at least half of the supply pressure.

Because of standard hose lengths, hydrants should not be located more than 300–400 feet (90–120 m) from the buildings they are intended to serve. Typical maximum spacings are 300 feet (90 m) for storage and distribution of petroleum oils, 350 feet (105 m) for normal uses, and 400 feet (120 m) for warehouses.

Fire hydrants should not be spaced much more than every 300 feet or 400 feet (90 m or 120 m). Hydrants should be spaced sufficiently close so that the hose stream demand can be satisfied without taking more than 750 gpm (2820 ℓ/min) from any single hydrant.

In order to be accessible for fire fighters, hydrants should be installed adjacent to paved areas, not closer than 3 feet (1 m) nor farther than 7 feet (2 m) from the roadway shoulder or curb line. The pumper connection should face (i.e., be perpendicular to) the street served.

Private Fire Service Mains (NFPA 24) and *Hose Connection Threads* (NFPA 1963) should be consulted regarding local fire hydrants.

Example 1

The capacity of a water supply tank for a schedule-designed fire sprinkler system is being evaluated. The building being protected is an ordinary hazard group 1 occupancy. The municipal water system does not have sufficient capacity to adequately supply the system. What should be the minimum tank capacity?

(Traditional U.S. Solution)

From Table 7 (see Sec. 52), the total water demand (including an allowance for hose streams) could be as high as 1500 gpm (95 ℓ/s).

Also from Table 7, the minimum duration is 60 minutes. The minimum required tank size is

$$V = t\dot{V} = (60 \text{ min})(1500 \text{ gal/min})$$
$$= 90{,}000 \text{ gal}$$

(SI Solution)

$$V = t\dot{V} = (60 \text{ min})(60 \text{ s/min})(95 \text{ } \ell/\text{s})$$
$$= 342\,000 \text{ } \ell$$

57. Supply Pumps for Fire Fighting[18]

Water from tanks and reservoirs usually flows under gravity action to its fire protection systems. However, depending on the elevations, supply pumps may be required. In some cases, the municipal system can supply adequate volume but at too low a pressure. In such cases, booster pumps can be used to boost the pressure of the municipal water supply. All pumps used in fire protection systems must be approved and/or listed. When a single pump constitutes the sole supply to sprinklers, it must be supervised (NFPA 13, Sec. 7-2.2.2).

Approved fire pumps are almost always horizontal (centrifugal) pumps. However, vertical (centrifugal or turbine) pumps are also used depending on which is more economical and appropriate. Pumps may be single or multiple stage. Horizontal pumps must operate with positive suction head. This is particularly important with remote starting. Horizontal pumps must be split-case, end-suction, or in-line types. Single-stage, end-suction, and in-line pumps are limited to under 500 gpm (NFPA 20, Sec. 3-1.1).

To avoid loss of prime and other suction lift problems, *vertical submersible turbine pumps* (i.e., "sump pumps") with submerged impellers can be used when drawing water from a deep well or sump. Vertical pumps can operate without priming, and they are used in ponds, streams, and pits. However, the elevation difference between the source (when pumping at 150% of rated

---

[18]NFPA 20, *Centrifugal Fire Pumps*, is the accepted authority on this subject.

capacity) and the ground surface is limited to 200 feet (NFPA 20, Sec. 4-1.2).

Pumps may be driven by diesel engines, but electric drives are preferred because of their simplicity. Electric power must be available from a reliable source (i.e., one experiencing fewer than two outages per year) or from two independent sources. Spark-ignited gasoline-powered engines should not be used.

Pumps should start automatically unless other sources are capable of simultaneously supplying water for all firefighting and industrial demands. Each pump should have a manual shutdown switch.

A *pressure-maintenance pump* (also known as a *jockey pump* or a *makeup pump*) is usually a low-volume, electrically-driven centrifugal pump. The impeller discharge is directly into the water sprinkler line. Operation is triggered by the initial water pressure drop resulting from the opening of a few sprinklers or other leakage. The automatic controller starts the primary fire pump when the capacity of the pressure-maintenance pump is exceeded.

There are three limiting points on the pump head-discharge curve:

1. *churn* or *shutoff*, where the pump operates at rated speed with the discharge valve closed

2. *rated (100%) capacity* and *rated (100%) head*

3. *overload,* 65% of the rated head

In selecting horizontal and vertical shaft turbine-type pumps for fire supply use, the churn (shutoff) head must not be greater than 140% of the rated head (NFPA 20, Secs. 3-2.1 and 4-1.3). This requirement eliminates pumps with high shutoff pressures and steep curves. The capacity at overload should not be less than 150% of the rated capacity (NFPA 20, Secs. 3-2.1 and 4-1.3). Figure 5 illustrates the pump characteristics that are described by these requirements.

Standard rated capacities of approved horizontal and vertical fire pumps are 25, 50, 100, 150, 200, 250, 300, 400, 450, 500, 750, 1000, 1250, 1500, 2000, 2500, 3000, 3500, 4000, 4500, and 5000 gpm (95, 189, 379, 568, 757, 946, 1136, 1514, 1703, 1892, 2839, 3785, 4731, 5677, 7570, 9462, 11 355, 13 247, 15 140, 17 032 and 18 925 ℓ/min). Pumps are rated at net pressures in excess of 40 psi (280 kPa; 2.8 bars) (NFPA 20, Sec. 2-3.1).

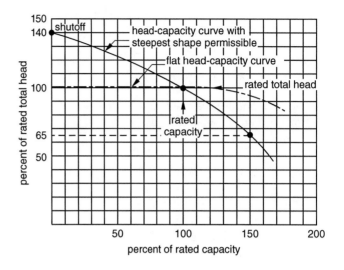

Figure 5 Standard Fire Protection Pump Curve

Reprinted with permission from NFPA 20, *Installation of Centrifugal Fire Pumps*, copyright © 1990, National Fire Protection Association, Quincy, MA 02269. This reprinted material is not the complete and official position of the National Fire Protection Association on the referenced subject, which is represented only by the standard in its entirety.

Pressure ratings range from (approximately) 40–200 psi (280 kPA–1.40 MPa; 2.8–14 bars) for horizontal pumps and from 75–280 psi (520 kPA–1.9 MPa; 5.2–19 bars) for vertical pumps. Pressure ratings include the pressure boost of all stages in multistage pumps.

Pumps that take suction from a private fire service main and supply only sprinklers (no hoses) do not have to be sized to meet inside and outside hose allowances (NFPA 13, Sec. 5-2.3.1.1, Exception 3).

It is general practice to choose a pump for fire protection systems based on overload capacity. This is illustrated in Ex. 2. However, some local authorities may require a pump to be sized below overload (i.e., with a higher capacity).

Example 2

A fire protection system requires 1300 gpm (82 ℓ/s), including hose streams of water at 80 psi (550 kPa). A suction tank is the source of the water. The equivalent pressure corresponding to the suction lift is 4 psi (28 kPa; 0.28 bars). What are the capacity and pressure ratings for an appropriate pump?

(Traditional U.S. Solution)

Meet the 1300 gpm demand with the pump's overload (i.e., 150%) capacity. A trial rated capacity is

$$Q = \frac{1300 \text{ gpm}}{1.50} = 867 \text{ gpm}$$

The nearest standard pump rating is 1000 gpm. Therefore, 1300 gpm would be 130% of capacity.

From Fig. 5, at 130% of capacity the total pressure is 80% of the rated pressure.

The total pump pressure is divided between the 80-psi sprinkler and 4-psi suction lift pressures. The total net pressure is

$$p_{net} = 80 \text{ psi} + 4 \text{ psi} = 84 \text{ psi}$$

The rated pressure at 1000 gpm is

$$p_{rated} = \frac{84 \text{ psi}}{0.80} = 105 \text{ psi}$$

The pump rating should be 1000 gpm at 105 psi.

(SI Solution)

Meet the 82 ℓ/s demand with the pump's overload (i.e., 150%) capacity. A trial rated capacity is

$$Q = \frac{82 \text{ } \ell/s}{1.50} = 54.7 \text{ } \ell/s$$

The nearest standard pump rating is 63 ℓ/s.

$$\frac{82 \text{ } \ell/s}{63 \text{ } \ell/s} = 1.30$$

Therefore, 82 ℓ/s would be 130% of capacity.

From Fig. 5, at 130% of capacity the total pressure is 80% of the rated pressure.

The total pump pressure is divided between the 550-kPa sprinkler and 28-kPa suction lift pressures. The total net pressure is

$$p_{net} = 550 \text{ kPa} + 28 \text{ kPa} = 578 \text{ kPa}$$

The rated pressure at 63 ℓ/s is

$$p_{rated} = \frac{578 \text{ kPa}}{0.80} = 723 \text{ kPa}$$

The pump rating should be 63 ℓ/s at 723 kPa.

58. Sprinkler Head Characteristics

The most common sprinkler types are the *upright sprinkler* (which discharges water upward against the deflector), the *pendant sprinkler* (which discharges water downward against the deflector), and the *sidewall sprinkler* (which emits water in a one-quarter sphere spray away from the wall, with a small amount directed at

the wall.)[19] Figure 6 illustrates the uniform water distribution pattern that is characteristic of sprinklers in use since 1953.

Several types of sprinklers have been used in the past including fusible link, glass bulb, and soldier puck sprinklers. Modern sprinklers come in a variety of designs, but all sprinklers contain several important elements: a nozzle, a deflector, and a release mechanism. The release mechanism for the old-type *fusible-link* (also known as soldered *link-and-lever) sprinkler* is a soldered fusible link that separates at a specific temperature. Figure 7 illustrates a link-and-lever sprinkler.

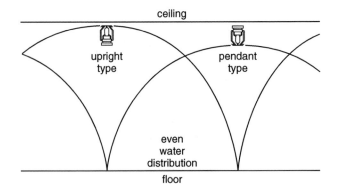

Figure 6　Distribution Pattern from Ceiling Sprinklers

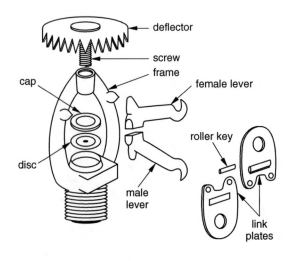

Figure 7　Soldered Link-and-Lever Automatic Sprinkler

[19]Older (i.e., pre-1953) sprinklers that discharge a fraction (i.e., 40%) of the water against the ceiling are no longer used except for special cases such as wharves and other wood construction.

59. Sprinkler Temperature Ranges

Sprinklers are assigned temperature classifications based on the temperature at which they are expected to operate. Frame arms of the sprinklers are color coded to visually indicate the classification. Table 8 summarizes the classifications. Other rated sprinklers may be available on special order.

Table 8
Temperature Ratings of Sprinklers

type	rated temperature range	maximum ceiling temperature	frame color
ordinary temperature	135, 150, 160, 165, 170°F (57, 66, 71, 74, 77°C)	100°F (38°C)	none[a]
intermediate temperature	175, 212, 225°F (79, 100, 107°C)	150°F (66°C)	white
high temperature	250, 280, 285, 300°F (121, 134, 141, 149°C)	225°F (107°C)	blue
extra-high temperature	325, 340, 350, 360, 375°F (163, 171, 177, 182, 191°C)	300°F (149°C)	red
very extra-high temperature	400–415°F (204–246°C)	375°F (191°C)	green
ultra-high temperature	500–575°F (260–302°C)	425°F (218°C)	orange
ultra-high temperature	650°F (343°C)	625°F (329°C)	orange

[a]Some manufacturers use half black/half uncolored to designate 135°F (57°C) sprinklers.

Reprinted with permission from NFPA 13, *Standard for the Installation of Sprinkler Systems*, copyright © 1991, National Fire Protection Association, Quincy, MA 02269. This reprinted material is not the complete and official position of the National Fire Protection Association on the referenced subject, which is represented only by the standard in its entirety.

The choice of a temperature rating depends on more than the normal ambient temperature in the protected space. Ratings are also chosen on the basis of occupancy/hazard type, number of heads intended to operate, proximity to other building features, and other characteristics.

Ordinary temperature-rated sprinklers are used unless ceiling temperatures exceed 100°F (38°C) (NFPA 13, Sec. 4-3.1.3). Ordinary temperature-rated sprinklers are also used in ventilated attics; under ventilated, peaked, and flat metal or wood roofs, whether insulated or not; and in ventilated show windows (NFPA 13, Table 4-3.2.3.2(b)).

Intermediate temperature-rated sprinklers are used near uncovered steam mains, heating coils, radiators, and glass and plastic skylights exposed to direct sunlight. They are also used in unventilated show windows illuminated by high-power electric lighting; in unventilated, concealed spaces and attics; and under uninsulated roofs.

High temperature-rated sprinklers are used in some warehouse applications (see *General Storage*, NFPA 231, and *Rack Storage of Materials*, NFPA 231C) and in heater zones, within 7 feet (2.1 m) of the free discharge of a low-pressure blow-off valve, to protect commercial cooking equipment (NFPA 13, 4-3.1.3.2).[20]

60. Sprinkler Life

For most sprinklers, the expected period of service (i.e., sprinkler life) will be approximately 50 years.[21] However, there are two exceptions:

1. High-temperature sprinklers (i.e., those rated above 325°F (163°C)) that have been exposed on a regular basis to maximum allowable temperatures should be tested every 5 years.

2. Residential and quick-response sprinklers (see Sec. 43) should be tested after 20 years of service and every 10 years thereafter.

Unopened sprinklers that have been exposed to a fire should be replaced (NFPA 13, Sec. 9-1-2).

61. Sprinkler Discharge Characteristics

The discharge from a sprinkler depends on the normal (i.e., static) pressure at the nozzle entrance. The minimum design pressure depends on the sprinkler head design and required coverage. Minimum pressure for any sprinkler is 7 psi (48 kPa; 0.48 bars), the pressure required to produce a 15-gpm flow (0.95 ℓ/s; 57 ℓ/min) in a standard $\frac{1}{2}$-inch-orifice (1.27-cm) sprinkler (NFPA 13, Sec. 6-4.4.8).

---

[20]Depending on the temperature, extra-high temperature-rated sprinklers may be used to protect commercial cooking equipment (NFPA 13, 4-3.1.3.2(g)).

[21]*Standard for the Installation of Sprinkler Systems* (NFPA 13) recommends replacing all sprinklers after 50 years of service.

If the *orifice constant* (also known as *nozzle constant*, *orifice discharge coefficient*, or *K-factor*), K, is known, Eq. 2 determines the relationship between the flow rate, Q, in gpm (ℓ/s), and normal pressure, p, in psi (kPa),

$$Q = K\sqrt{p} \qquad \text{[U.S.] 2(a)}$$

$$Q_{\ell/s} = K_{\text{SI}}\sqrt{p} \qquad \text{[SI] 2(b)}$$

$$Q_{\ell/\text{min}} = K_{\text{bars}}\sqrt{p_{\text{bars}}} \quad [p \text{ in bars}] \qquad \text{2(c)}$$

$$K_{\text{SI}} = 0.024K \qquad \text{3(b)}$$

$$K_{\text{bars}} = 14K \qquad \text{3(c)}$$

Although there are many sprinkler designs, most have similar operating characteristics. For example, the standard heads have a 0.5-inch (1.27-cm) orifice, and they discharge 15 gpm (0.95 ℓ/s; 57 ℓ/min) at 7 psi (48 kPa; 0.48 bars). This is equivalent to a discharge constant, K, of approximately 5.6–5.8 (0.134–0.140). The coefficient of discharge, C_d, is taken as 0.75–0.78. Figure 8 illustrates the discharge characteristics for standard 0.5-inch (12.7-mm) and $\frac{17}{32}$-inch (13.5-mm) orifices.

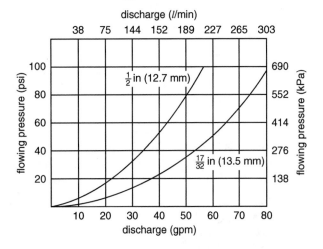

pressure at sprinkler psi (kPa)	approximate discharge gpm (l/min)	
	$\frac{1}{2}$ in (12.7 mm)	$\frac{17}{32}$ in (13.5 mm)
10 (69)	18 (68)	25 (95)
15 (103)	22 (83)	31 (117)
20 (138)	25 (95)	36 (136)
25 (172)	28 (106)	40 (151)
35 (241)	33 (125)	47 (178)
50 (345)	40 (151)	57 (216)
75 (517)	48 (182)	69 (261)
100 (690)	56 (212)	80 (303)

Figure 8 Orifice Discharge vs. Static Pressure
(0.5-in and $\frac{17}{32}$-in diameter orifices)

Reprinted with permission from *Fire Protection Handbook*, copyright © 1991, National Fire Protection Association, Quincy, MA 02269. This reprinted material is not the complete and official position of the National Fire Protection Association on the referenced subject, which is represented only by the standard in its entirety.

Heads with smaller orifices can be used for light hazard occupancies when hydraulically-designed. Strainers are required for orifices less than $\frac{3}{8}$ inch (9.5 mm) in diameter (NFPA 13, Sec. 2-2.2.1). Strainers are also a good idea with high-velocity flows. Strainers are often installed in risers.

Example 3

A sprinkler has a discharge constant of 5.6 (0.13 for liters and seconds). What is the discharge from the sprinkler if the normal pressure is 23 psi (160 kPa)?

(Traditional U.S. Solution)

$$Q = K\sqrt{p} = (5.6)\sqrt{23 \text{ psi}}$$
$$= 26.9 \text{ gpm}$$

(SI Solution)

$$Q = K\sqrt{p} = (0.13)\sqrt{160 \text{ kPa}}$$
$$= 1.64 \ \ell/s$$

Table 9
Sprinkler Discharge Characteristics

nominal orifice size (in)	orifice type	K-factor	percent of nominal $\frac{1}{2}$-in discharge
$\frac{1}{4}$	small	1.3–1.5	25
$\frac{5}{16}$	small	1.8–2	33.3
$\frac{3}{8}$	small	2.6–2.9	50
$\frac{7}{16}$	small	4.0–4.4	75
$\frac{1}{2}$	standard	5.3–5.8	100
$\frac{17}{32}$	large	7.4–8.2	140
$\frac{5}{8}$	extra large	11.0–11.5	200
$\frac{5}{8}$	large drop	11.0–11.5	200
$\frac{3}{4}$	ESFR	13.5–14.5	250

Reprinted with permission from NFPA 13, *Standard for the Installation of Sprinkler Systems*, copyright © 1991, National Fire Protection Association, Quincy, MA 02269. This reprinted material is not the complete and official position of the National Fire Protection Association on the referenced subject, which is represented only by the standard in its entirety.

62. Sprinkler Protection Area

Theoretical coverage per sprinkler can vary from 60–400 ft² (5.4–36 m²) of floor area. However, the common operating range is more in the order of 80–120 ft² (7.2–10.8 m²) of floor area.

The spacing and layout design of sprinkler systems will be governed by the maximum floor area that one sprinkler can be expected to protect (i.e., the *protection area*). This, according to *Standard for the Installation of Sprinkler Systems* (NFPA 13), is in turn determined by two factors: the hazard classification and the type of ceiling construction. (Sidewall spray sprinklers have their own different area and spacing requirements (NFPA 13, Sec. 4-4.2).) Other sprinklers may be listed for larger coverage areas.

For example, light hazard areas (Sec. 3) as listed in Table 10 require one sprinkler for every 200 ft^2 (18 m^2) of floor area if the design is by schedule. The requirement is relaxed to one sprinkler for every 225 ft^2 (20.3 m^2) if the system is hydraulically calculated. For open wood joist ceilings, one sprinkler is required for every 130 square feet (12 m^2) (NFPA 13, Sec. 4-2.2).

Table 10
Maximum Sprinkler Protection Areas

	light hazard	ordinary hazard	extra hazard[e]	high-piled storage[f]
unobstructed ceiling[a]	225 ft^2[b]	130 ft^2	100 ft^2	100 ft^2
noncombustible obstructed ceiling	225 ft^2[b]	130 ft^2	100 ft^2	100 ft^2
combustible obstructed ceiling	168 ft^2[c][d]	130 ft^2	100 ft^2	100 ft^2

Multiply ft^2 by 0.0929 to obtain m^2.
Multiply gpm/ft^2 by 40.75 to obtain ℓ/min·m^2.

[a] Unobstructed construction excludes wood truss construction.

[b] 200 ft^2 (18 m^2) when designed by schedule.

[c] 130 ft^2 (11.7 m^2) for light combustible framing members spaced less than 3 ft on center.

[d] 225 ft^2 (22.3 m^2) for heavy framing members spaced 3 ft or more on center.

[e] 90 ft^2 when designed by schedule. 130 ft^2 (11.7 m^2) when hydraulically designed and density is less than 0.25 gpm/ft^2 (10.2 ℓ/min· m^2).

[f] 130 ft^2 (11.7 m^2) when hydraulically designed according to NFPA 231 and NFPA 231C and density is less than 0.25 gpm/ft^2 (10.2 ℓ/min·m^2).

Reprinted with permission from NFPA 13, *Standard for the Installation of Sprinkler Systems*, copyright © 1991, National Fire Protection Association, Quincy, MA 02269. This reprinted material is not the complete and official position of the National Fire Protection Association on the referenced subject, which is represented only by the standard in its entirety.

When calculating numbers of sprinklers required for areas by dividing the floor area by the sprinkler coverage, fractional numbers are rounded up to the next higher integer. Similarly, when calculating numbers of sprinklers required for branches by dividing the design length by the sprinkler spacing, fractional numbers are rounded up to the next higher integer.

The hydraulically most-demanding area may not be rectangular when determined from the number of sprinklers, and one or more extra sprinklers may need to be included from an adjacent line. (This is the case in Ex. 4.) For gridded systems, the extra sprinklers can be any adjacent sprinkler, at the designer's option. For tree and looped systems, the extra sprinkler should be the one adjacent sprinkler closest to the cross main (NFPA 13, App. A-6-4.4).

Even though the number of sprinklers is determined assuming that all sprinklers will discharge at the same rate, this does not actually occur in practice. The discharge rate depends on the pressure at the nozzle, which decreases toward the last open sprinkler. This introduces a conservative element in the design. The first (by time) sprinkler to open will discharge at a higher-than-average design rate until other sprinklers open.

63. Discharge Density

Another design criterion, increasingly used in severe hazard areas (including high-piled storage and combustible liquid storage), calls for a discharge density in gpm/ft^2 (ℓ/m^2· s). Values between 0.15 and 0.6 gpm/ft^2 (0.10–0.40 ℓ/m^2· s) are typical.

Sizing supply lines using discharge density design criteria is known as the *area/density method*. When specified by code or building officials, the area/density method takes precedence over other design criteria (NFPA 13, Sec. 5-2.3.1.1, Exception No. 1).

In order to determine the discharge density, the *design area* must first be known. This can be specified by local ordinances, the local fire or building officials, insurance requirements, or the fire protection engineer. The design area is not the same as the coverage per sprinkler or the total room/building area. The total discharge from sprinklers in the design area is added to the hose allowance to determine the required water supply.

Figure 9 contains area/density curves for different occupancy types. Subsequent calculations must satisfy any one point on the appropriate curve. It is not necessary to meet all points of the appropriate curve.

The area/density curves illustrated in Fig. 9 indicate that larger design areas have lower densities. This may result in inadequate protection for smaller areas. Therefore, the density corresponding to a design area of

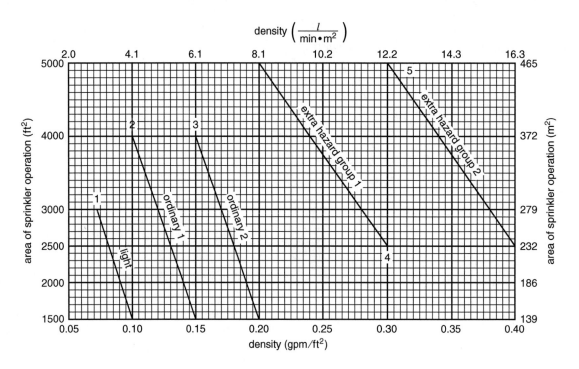

density $\left(\frac{l}{\min \cdot m^2}\right)$

Multiply ft^2 by 0.0929 to obtain m^2.
Multiply gpm/ft^2 by 40.75 to obtain ℓ/min·m^2.

Figure 9   Area/Density Discharge Curves for Wet Spray Systems[a],[b] (NFPA 13, Fig. 5-2.3)

[a] NFPA 13 has other criteria for sprinklers other than spray (NFPA 13, Secs. 5-2.3.2.2 and 5-2.3.2.4).

[b] For dry pipe systems, the area is increased by 30% without increasing the density (NFPA 13, Sec. 5-2.3.2.3).

Reprinted with permission from NFPA 13, *Standard for the Installation of Sprinkler Systems*, copyright © 1991, National Fire Protection Association, Quincy, MA 02269. This reprinted material is not the complete and official position of the National Fire Association on the referenced subject, which is represented only by the standard in its entirety.

1500 ft^2 (139 m^2) is used for light and ordinary hazard occupancies when the actual area being protected is less than 1500 ft^2 (139 m^2). Similarly, the density corresponding to a design area of 2500 ft^2 (232 m^2) is used for extra hazard occupancies when the actual area is less than 2500 ft^2 (232 m^2) (NFPA 13, Sec. 5-2.3.1.3).

The discharge density affects the sprinkler system design by establishing the discharge from the hydraulically most-distant sprinkler. (The discharge is the coverage area per sprinkler multiplied by the discharge density.)

The discharge, in turn, establishes (see Eq. 2) the pressure at the most-distant sprinkler. The discharge from sprinklers closer to the cross-main is calculated from the available pressure, not the discharge density.

When the area/density method is used, the design area is assumed to be rectangular, with a dimension parallel to the branch lines (perpendicular to the cross-main) of at least 1.2 times the square root of the actual sprinkler operation area, A. (This value may be changed from 1.2 to 1.4 by the local authorities or the insurance underwriter.) This can include area on both sides of the cross-main. If the actual branch lines are too short to fulfill the $1.2\sqrt{A}$ requirement, the design area is extended down the main to include adjacent branches (NFPA 13, Sec. 6-4.4.1(a)).

Example 4

An ordinary occupancy (group 1) building, 20 feet high, 140 feet wide, and 140 feet long, has a ceiling-beam spacing of 20 feet. The beams run the full width of the building, dividing the building into 20-foot bays. The beams block water spray from sprinklers in adjacent bays. Sprinklers will be placed every 12 feet along branch lines. The local building official requires a sprinkler design area of 1500 ft^2. (a) Design the sprinkler system branches. (b) Draw the design area for purpose of calculating the required water supply. (c) Assuming an average sprinkler discharge of 22 gpm and an allowance for hose streams of 250 gpm, what is the total water supply required?

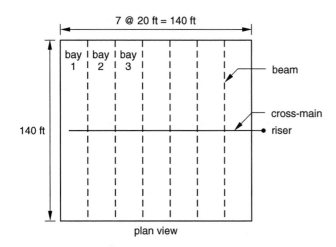

plan view

(Solution)

(a) The distance between beams is greater than 15 feet (the maximum spacing between sprinklers), so two branch lines are required. Space the two branches 10 feet apart.

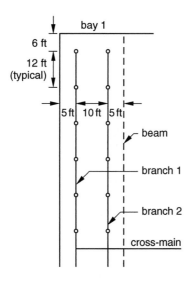

(b) Since the distance between sprinklers along branch lines is 12 feet, the coverage per sprinkler is

$$A_{\text{sprinkler}} = (10\,\text{ft})(12\,\text{ft}) = 120\,\text{ft}^2$$

The number of sprinklers in the design area is

$$\frac{1500\,\text{ft}^2}{120\,\text{ft}^2} = 12.5 \quad [\text{use 13}]$$

The length of the design area along each branch is

$$1.2\sqrt{A} = 1.2\sqrt{1500\,\text{ft}^2} = 46.5\,\text{ft}$$

The number of sprinklers on each branch line in the design area is

$$\frac{46.5\,\text{ft}}{12\,\text{ft}} = 3.88 \quad [\text{use 4}]$$

The number of branches in the design area is

$$\frac{13}{4} = 3.25\ \text{branches}$$

There are only two branches in each bay, so the design area is extended to include one entire branch in the adjacent bay and one additional sprinkler. This is a tree system, so the additional sprinkler must be D, the sprinkler closest to the cross-main. The design area is shown.

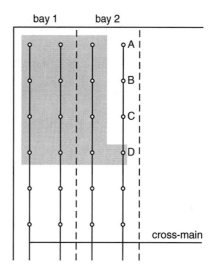

(c) 13 sprinklers are in the design area. The total sprinkler flow is

$$(13)(22\,\text{gpm}) = 286\,\text{gpm}$$

Including a 250-gpm allowance for hose streams, the water supply requirement is

$$286\,\text{gpm} + 250\,\text{gpm} = 536\,\text{gpm}$$

64. Distance Between Sprinklers

The distance between sprinklers on lines should be approximately the same as the distance between lines. The maximum spacing between adjacent sprinkler heads and branch lines is generally 15 feet (4.5 m) for light and ordinary hazard occupancies and 12 feet (3.6 m) for extra hazard and high-piled storage occupancies. There are two rare exceptions to these spacing requirements for bays and low discharge densities (NFPA 13, Sec. 4-4.1.1). (Sidewall spray sprinklers have their own, different area and spacing requirements (NFPA 13, Sec. 4-4.2).)

The maximum distance to the nearest wall is generally one-half of the required spacing but not less than 4 inches (10.2 cm) (NFPA 13, Sec. 4-4.1.2.1).

Sprinklers that are spaced too close together may interfere with other sprinklers' openings. For example, the water discharge from one sprinkler may cool and delay or prevent an adjacent sprinkler from opening. If sprinklers are spaced less than every 6 feet (1.8 m), baffles are required between them (NFPA 13, Sec. 4-4.1.7.8). This limitation might be increased to 8 feet by the local authority or the insurance underwriter. NFPA 13 should be consulted for baffle requirements for other types of sprinklers, such as large drop and early suppression fast response (ESFR) sprinklers.

Structural members (including joists and trusses) and nonflat ceiling construction greatly complicates spacing and placement standards. NFPA 13, Sec. 4-4.1.3 should be consulted in such instances.

65. Ceiling Height

Since air temperature increases as the distance from the floor increases, flush-mounted ceiling sprinklers will open before dropped sprinklers. On one hand, sprinklers should be as close to the ceiling as possible in order to minimize response time. On the other hand, the lateral distribution pattern may be adversely affected (except in plane ceilings) by girders and beams. *Standard for the Installation of Sprinkler Systems* (NFPA 13) specifies ceiling distances for different ceiling configurations.

66. Protecting Process Tanks

Special requirements apply to specific hazard elements. For example, chemical reaction vessels, process tanks, and structural steel members may be given their own spray systems.

Uninsulated steel vessels under average plant conditions can be expected to absorb heat at the rate of approximately 20,000 BTU/ft^2-hr (63 kW/m^2) of exposed surface wetted internally by the vessel contents.[22] Applying a water spray will reduce the heat input rate to approximately 6000 BTU/ft^2-hr (19 kW/m^2) of internally wetted area when 0.2 gpm/ft^2 (0.14 ℓ/m$^2\cdot$s) of

water is applied. Limited experimental test data indicates that the water application rate should be approximately 0.60 gpm/ft^2 (0.42 ℓ/m$^2\cdot$s) when a leaking chemical tank contains flammable gas.

67. Schedule Design

The applicability of schedule sprinkler system design was discussed in Sec. 41. Table 11 is a typical schedule of the maximum number of sprinklers heads that can be fed by a pipe of a given size. The number of sprinkler heads includes all downstream sprinklers, no matter what size pipe they are actually installed in. Table 11 gives the maximum number of sprinklers that can be supported on pipes of various sizes for light and ordinary hazards. Sprinklers for new extra hazard occupancies must be hydraulically calculated (NFPA 13, Sec. 6-5.4).[23]

Table 11
Typical Schedule of
Sprinklers vs. Pipe Size
(wet and dry systems; spacing 12 ft or less[a])

pipe size (in)	maximum number of sprinklers[b]	
	light hazard	ordinary hazard
$\frac{3}{4}$	none	none
1	2	2
$1\frac{1}{4}$	3	3
$1\frac{1}{2}$	5	5
2	10 (12)	10 (12)
$2\frac{1}{2}$	30 (40)	20 (25)
3	60 (65)	40 (45)
$3\frac{1}{2}$	100 (115)	65 (75)
4	no limit	100 (115)
5	no limit	160 (180)
6	no limit	275 (300)

[a] See Sec. 69 for values when sprinkler spacing is between 12 and 15 ft.

[b] Numbers in parentheses are for copper pipe.

Reprinted with permission from NFPA 13, *Standard for the Installation of Sprinkler Systems*, copyright © 1991, National Fire Protection Association, Quincy, MA 02269. This reprinted material is not the complete and official position of the National Fire Protection Association on the referenced subject, which is represented only by the standard in its entirety.

---

[22] Vessel walls not wetted by contents absorb heat at a much greater rate.

[23] Schedule design for extra hazard design was at one time permitted. Refer to NFPA 13, App. A-6-5.4 for the pipe schedule to be used with existing extra hazard schedule-designed systems.

The numbers of sprinklers in Table 11 should be reduced when there are long runs of pipe or many bends and fittings (NFPA 13, App. A-6-5.1.2). In the absence of hydraulic calculations, the reduction is a matter of engineering judgment.

When a large number of heads are expected to be open simultaneously (e.g., deluge systems and other hazardous environments), the maximum number of sprinkler heads should be reduced. This is also a matter of judgment.

68. Maximum Number of Branch Sprinklers

When schedule design is used, limits are also placed on the maximum number of sprinklers that can be placed on branch lines on either side of a cross-main. For example, branch lines for light and ordinary hazard occupancies cannot have more than eight sprinklers on either side of the cross-main; for extra hazard occupancies, the limitation is six (NFPA 13, Secs. 6-5.2.1 and 6-5.3.1). Other restrictions apply when sprinklers are installed both above and below a ceiling.[24]

69. Special Layout Rules

There are many special layout rules that are observed. Two of the most commonly-used are presented here.

1. Where the layout is a corridor protected by a single row of sprinklers, the maximum number of sprinklers that needs to be calculated is 5 (NFPA 13, Sec. 5-2.3.4.2).

2. When the spacing between sprinklers is greater than 12 feet but less than or equal to 15 feet, Table 12 gives the maximum number of sprinklers on a branch.

Table 12
Maximum Number of Sprinklers
(ordinary hazard; spacing greater than 12 ft)
(NFPA 13, Table 6-5.3.2(b))

pipe size (in)	steel pipe	copper
$2\frac{1}{2}$	15	20
3	30	35
$3\frac{1}{2}$	60	65

Reprinted with permission from NFPA 13, *Standard for the Installation of Sprinkler Systems*, copyright © 1991, National Fire Protection Association, Quincy, MA 02269. This reprinted material is not the complete and official position of the National Fire Protection Association on the referenced subject, which is represented only by the standard in its entirety.

---

[24]There are special cases, exceptions, and compensating methods for these restrictions. NFPA 13 should be consulted for these situations.

Example 5

A remote square-shaped room has an actual area of 1800 ft². Sprinklers with a 130 ft² coverage are used with a maximum 12-foot spacing. How many sprinklers are used on each branch line?

(Solution)

(This problem specifies actual room area, not a design area.)

The total number of sprinklers is

$$n = \frac{1800 \text{ ft}^2}{130 \text{ ft}^2} = 13.85 \quad [14 \text{ sprinklers minimum}]$$

Calculate the number of sprinklers on each branch line. The area dimension is

$$\sqrt{1800 \text{ ft}^2} = 42.4 \text{ ft}$$

Since the maximum spacing is 12 feet, the minimum number of sprinklers on each branch line is

$$\frac{42.4 \text{ ft}}{12 \text{ ft}} = 3.5 \text{ sprinklers/branch}$$

This is rounded up to 4 sprinklers per branch line.

The sprinkler layout is developed by trial and error. Installing more sprinklers than is required by the coverage calculation is usually inevitable. The distances shown in the layout add up to 42 feet. The extra 0.4 feet can be added at any convenient location. Spacings other than even-foot can be used at the designer's option.

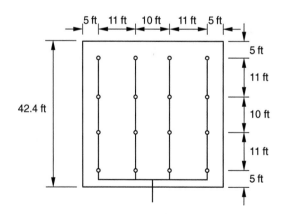

70. Area Covered by Sprinklers

In analyzing an actual or perspective layout, the area covered by any sprinkler along a branch line is traditionally determined according to the following procedure (NFPA 13, Sec. 4-2.2.1.1).

step 1: Determine the maximum of (a) the distance from the given sprinkler to the next sprinkler on the branch line and (b) for end sprinklers, double the distance from the given sprinkler to the wall or obstruction. Call this distance S.

step 2: Determine the maximum of (a) the distance from the branch line containing the given sprinkler to the adjacent, parallel branch line and (b) for end sprinklers, double the distance from the given sprinkler to the wall. Call this distance L.

step 3: The design area covered by the sprinkler is

$$A = SL \qquad\qquad 4$$

Example 6

Determine the maximum area covered by a sprinkler in the final design of Ex. 5.

(Solution)

The maximum distance from one sprinkler to the next along the same branch is $S = 11$ feet. The maximum distance between branches is $L = 11$ feet. Both of these are more than twice the 5-foot sprinkler-to-wall spacing.

The maximum area covered by a sprinkler is

$$A = SL = (11\text{ ft})(11\text{ ft}) = 121\text{ ft}^2$$

71. Friction Losses

The Hazen-Williams equation (Eq. 5) should be used to calculate the friction loss in hydraulically-designed sprinkler systems. C is the Hazen-Williams coefficient, commonly known as the C-value. The actual (not nominal) pipe diameter should be used with Eq. 5.

$$p_{f,\text{psi}} = \frac{4.52 L_{\text{ft}} Q_{\text{gpm}}^{1.85}}{C^{1.85} d_{\text{in}}^{4.87}} \qquad \text{[U.S.] } 5(a)$$

$$p_{f,\text{kPa}} = \frac{(1.18 \times 10^8) L_{\text{m}} Q_{\ell/\text{s}}^{1.85}}{C^{1.85} d_{\text{mm}}^{4.87}} \qquad \text{[SI] } 5(b)$$

Use Eq. 5(c) to calculate a pressure drop in bars from a flow rate in ℓ/min.

$$p_{f,\text{bars}} = \frac{(6.05 \times 10^5) L_{\text{m}} Q_{\ell/\text{min}}^{1.85}}{C^{1.85} d_{\text{mm}}^{4.87}} \qquad 5(c)$$

Table 13 gives the Hazen-Williams C-values to be used. The local building or fire officials, however, may specify

other values. The C-value is assumed to be constant for a specific pipe roughness and is independent of velocity.

Table 13
Hazen-Williams C-Values
for Sprinkler System Design

type of pipe	C
unlined cast or ductile iron	100
black steel (for dry and preaction systems)	100
black steel (for wet and deluge systems)	120
galvanized (all uses)	120
plastic (listed)	150
cement-lined cast or ductile iron	140
copper tube or stainless steel	150

New unlined steel pipe has a C-value of 140. However, as the pipe ages, the friction will increase. For that reason, 120 is the value used for wet systems. Dry systems have an even greater tendency to develop deposits and corrosion. Therefore, the C-value for dry systems is taken as 100. Both of these assumptions are conservative. When the pipe is new, greater-than-design flows will be achieved.

Graphical solutions to the Hazen-Williams equation are permitted and are commonly used. Appendices B and C are graphical solutions for $C = 120$. Friction losses for other C-values can be calculated by multiplying the graphical friction loss by the appropriate factor from Table 14.

Table 14
Multipliers for Other C-Values

	basis of graphical value	
actual C-value	$C = 100$	$C = 120$
150	0.472	0.662
140	0.537	0.752
130	0.615	0.862
120	0.714	1.00
110	0.838	1.18
100	1.00	1.40
90	1.22	1.70
80	1.51	2.12
70	1.93	2.71
60	2.57	3.61

72. Minor Losses

Ideally, all sources of friction, bends, valves, meters, and strainers should be recognized. However, in hydraulic calculations for sprinkler systems, only fittings involving changes in flow direction are included (NFPA 13,

Sec. 6-4.4.5(b)). Minor losses from straight-through run-of-tees and straight-across crosses are disregarded. Friction losses due to tapered reducers and reducers directly adjacent (attached) to spray nozzles are also disregarded.

Minor losses from tees at the top of risers are included with the cross-mains, losses for tees at the bases of risers are included with the risers, and losses from crosses or tees at cross-main feed junctions are included with the cross-mains (NFPA 13, Sec. 6-4.4.5(b)).

For tees and reducing elbows, the equivalent length is based on the velocity and/or diameter of the smaller outlet (NFPA 13, Sec. 6-4.4.5(c)).

Values for standard elbows are typically used for abrupt 90° turns constructed with threaded fittings. The long-turn elbow values are used with flanged, welded, or other mechanical connections (NFPA 13, Sec. 6-4.4.5(c)).

Friction losses for fittings connected directly to sprinklers are omitted (NFPA 13, Sec. 6-4.4.5(d)).

The equivalent lengths of valves and fittings for $C = 120$ are given in App. D. Loss data for specialized elements (such as pressure-reducing valves, deluge valves, alarm valves, dry pipe valves, and strainers) must be obtained from the manufacturers (NFPA 13, Secs. 6-4.3.3, 6-4.4.5(e)).

Friction losses for other C-values can be calculated by multiplying the equivalent lengths in App. D by the appropriate factor in Table 15. (These factors assume that the friction loss through the fitting is independent of the piping C-value (NFPA 13, Sec. 6-4.3.2).)

Table 15
Multiplying Factors for Equivalent Lengths

actual C-value	multiplying factor
100	0.713
120	1.00
130	1.16
140	1.33
150	1.51

73. Velocity Pressure

Velocity pressure, since it is small, may be omitted at the discretion of the designer. This normally introduces a conservative error. However, if the velocity pressure is much more than 5% of the total pressure, it should be considered. If considered, velocity pressure must be included in calculations for both nonlooped branch lines and cross-mains (NFPA 13, Sec. 6-4.4.7).

$$p_{v,\text{psi}} = \frac{0.433\, v^2}{2g} \qquad \text{[U.S.]} \quad 6(a)$$

$$p_{v,\text{kPa}} = \frac{9.81\, v^2}{2g} \qquad \text{[SI]} \quad 6(b)$$

$$p_{v,\text{bars}} = \frac{0.0981\, v^2}{2g} \qquad 6(c)$$

Even when velocity pressure is considered, however, the method used to include it is peculiar to sprinkler design. Specifically, the velocity pressure downstream of a sprinkler is used for end outlets, while velocity pressure upstream (on the supply side) is used for other outlets (NFPA 13, App. A-6-4.4.6(b)). This strange convention is illustrated in Ex. 9.

End outlets include the last sprinkler on a dead-end branch, the final flowing branch on a dead-end cross-main, any sprinkler with a flow split on a gridded branch line, and any branch line with a flow split on a loop system (NFPA 13, App. A-6-4.4.7(a)).

When velocity pressure is based on the upstream flow rate (which is initially unknown), the velocity pressure is determined by trial and error. This procedure starts with an estimated flow rate for the upstream side of the nozzle. This flow rate is used to determine the trial velocity pressure, and in turn, the normal pressure and a new flow rate. The procedure is repeated until the estimated and calculated flow rates converge sufficiently.

In analysis problems that consider velocity pressure, the discharge from the next-to-last sprinkler in a line may be lower than the last sprinkler. This condition may or may not require attention. If the discharge density from the next-to-last sprinkler exceeds the minimum and is less than approximately 3% of the total design flow, the design probably should be kept. Otherwise, the pipe supplying the end sprinkler should be increased in size.

Example 7

The flow through a 1-inch (nominal) schedule-40 branch line is 36 gpm (2.3 ℓ/s). What is the velocity pressure?

(Traditional U.S. Solution)

The internal cross-sectional area of a 1-inch line is 0.0060 ft².

The velocity is

$$v = \frac{\dot{V}}{A} = \frac{(36\text{ gpm})\left(0.002228\ \dfrac{\text{ft}^3}{\text{sec-gpm}}\right)}{0.0060\text{ ft}^2}$$
$$= 13.37\text{ ft/sec}$$

From Eq. 6, the velocity pressure is

$$p_v = \frac{0.433\, v^2}{2g}$$

$$= \frac{\left(0.433\, \dfrac{\text{psi}}{\text{ft}}\right)\left(13.37\, \dfrac{\text{ft}}{\text{sec}}\right)^2}{(2)\left(32.2\, \dfrac{\text{ft}}{\text{sec}^2}\right)}$$

$$= 1.20\text{ psi}$$

(SI Solution)

The internal cross-sectional area of a 1-inch line is $5.574 \times 10^{-4}\ \text{m}^2$.

The velocity is

$$v = \frac{\dot{V}}{A} = \frac{\left(2.3\, \dfrac{\ell}{\text{s}}\right)\left(0.001\, \dfrac{\text{m}^3}{\ell}\right)}{5.574 \times 10^{-4}\ \text{m}^2}$$

$$= 4.13\text{ m/s}$$

From Eq. 6, the velocity pressure is

$$p_v = 9.81\, \frac{v^2}{2g}$$

$$= \frac{\left(9.81\, \dfrac{\text{kPa}}{\text{m}}\right)\left(4.13\, \dfrac{\text{m}}{\text{s}}\right)^2}{(2)\left(9.81\, \dfrac{\text{m}}{\text{s}^2}\right)}$$

$$= 8.53\text{ kPa}$$

74. Normal Pressure

The *normal pressure*, usually referred to in other subjects as the *static pressure*, is the difference between the total pressure and the velocity pressure.

$$p_n = p_t - p_v \qquad\qquad 7$$

75. Hydraulic Design Concepts

In order to design a sprinkler system according using industry-accepted methods, the following information must be known:

- area to be protected in square feet (m^2)
- minimum rate of water application (density) in gpm/ft^2 $(\ell/\text{min}\cdot\text{m}^2)$
- sprinkler information, such as coverage per sprinkler
- discharge constant for each nozzle type, or discharge curve
- pipe sizes

- layout and pipe lengths
- equivalent pipe lengths for fittings and other devices
- friction loss C-value, or friction loss per length of pipe
- elevation changes
- minimum acceptable normal pressure at sprinklers
- required pressure at reference points
- flow allowances for inside hose, outside hydrants, and in-rack sprinklers

Hydraulic calculations, whether for analysis or design, can be performed by hand or by computer. However, to standardize designs, certain conventions (which follow) have been established. This includes standardized terminology (i.e., *normal pressure* instead of *static pressure*) and nomenclature.[25] Alternatively, more sophisticated methods may be used. However, it may be more difficult to rationalize these methods to the local officials.

Analysis calculations of sprinkler systems start at the hydraulically most-remote nozzle using the minimum nozzle pressure (e.g., 20 psig). A worksheet similar to App. F must be used (NFPA 13, Sec. 6-2.1). This most-remote nozzle is not necessarily the most-distant nozzle. Rather, it is the nozzle whose supply-to-nozzle path experiences the greatest friction loss. It may take several trial sets of calculations to verify which sprinkler is the most remote hydraulically. The analysis procedure is detailed in Sec. 76.

To begin the design, either the pressure or the discharge quantity must be known for the most-remote nozzle. If the pressure is known, the discharge quantity is found from the pressure-volume curve for the sprinkler. If the density requirement is known, it can be used to calculate the discharge quantity. The pressure at the nozzle is then found from the pressure-volume curve. Calculations of normal pressure then proceed, fitting by fitting, back to the point of water supply.

When there is only one sprinkler, as there is for typical protection of a closet or washroom, a reasonable assumption is that the sprinkler will be independent of the other sprinklers in the main room. As long as the sole sprinkler can discharge at the necessary rate, it can be omitted from the hydraulic calculations for main areas greater than 1500 ft^2 (135 m^2) (NFPA 13, Sec. 6-4.4.4, Exception 1).

---

[25]The conventions and standardized nomenclature in this chapter are taken from *Standard for the Installation of Sprinkler Systems* (NFPA 13).

Branch calculations should be performed separately (i.e., on their own sheets). Calculations are normally omitted for identical and symmetrical branches.

Sprinkler systems supplied by loops are more difficult to analyze and design. Although looped systems can be designed by hand using the Hardy-Cross method, it is more expedient to use computerized methods.

76. Hydraulic Design Procedure: Pressure Along Branches

The following analysis procedure starts at the hydraulically most-distant sprinkler. Note the difference in how velocity pressure is calculated for the third-to-last sprinkler in a branch and beyond.

step 1: Calculate the discharge rate (in gpm) for the last sprinkler from the area and density or from the normal pressure and discharge characteristics, whichever is known. Either Eq. 2 or graphical characteristics can be used. (No sprinkler may discharge at less than the minimum design rate.)

step 2: Using the flow rate, pipe length to the next sprinkler, and C-value, determine the Hazen-Williams friction pressure loss (in psi). Either Eq. 5, App. B, or App. C may be used. If the Hazen-Williams equation is solved graphically, convert the loss to the value corresponding to the actual C-value.

step 3: Add the friction pressure loss from step 2 to the normal pressure from step 1. This is the total pressure available at the next-to-last sprinkler.

step 4: If it is to be included in the calculations, calculate the velocity pressure (in psi) in the section of pipe used in step 2 from Eq. 6.

step 5: Subtract the velocity pressure from the total pressure. This is the normal pressure available at the next-to-last sprinkler. Notice that the downstream, not upstream, velocity pressure is used with the next-to-last sprinkler.

step 6: Calculate the discharge from the next-to-last sprinkler from the normal pressure and sprinkler discharge characteristics. (No sprinkler may discharge at less than the minimum design rate.)

step 7: Using the cumulative flow rate and pipe length to the next sprinkler, determine the Hazen-Williams friction pressure loss (in psi). Convert pressure loss to proper C-value if necessary.

step 8: If velocity pressure is to be considered, calculate it from the flow rate downstream of the next sprinkler.

step 9: Calculate the total pressure at the next downstream sprinkler by adding the normal pressure at the previous sprinkler, the friction pressure from step 7, and the velocity pressure from step 8.

step 10: Estimate the sprinkler discharge for the next downstream sprinkler. (Finding the discharge from the third-to-last and all subsequent sprinklers is an iterative process. The upstream velocity is used with all sprinklers except the next-to-last. Since the upstream velocity is unknown, it must be estimated. While the estimated variable could be either the upstream velocity or upstream flow rate, it is common (and easier) to estimate the sprinkler discharge.)

step 11: Calculate the flow rate in the upstream pipeline by adding the estimated sprinkler discharge and the cumulative discharge from subsequent sprinklers.

step 12: Calculate the upstream velocity pressure from the flow rate determined in step 11.

step 13: Calculate the normal pressure by subtracting the velocity pressure calculated in step 12 from the total pressure calculated in step 9.

step 14: Determine a corrected discharge from the sprinkler using the normal pressure calculated in step 13. (No sprinkler may discharge at less than the minimum design rate.)

step 15: Compare the discharge assumed in step 9 with the corrected discharge calculated in step 14. If they are sufficiently the same, repeat steps 7 through 15 for all remaining sprinklers. If the two values are different, estimate a new discharge and repeat steps 10 through 15.

Example 8

The hydraulically most-remote branch of a sprinkler system consists of a 1-inch (nominal) schedule-40 steel pipe and three sprinklers with standard $\frac{1}{2}$-inch-diameter orifices. The last sprinkler is at the end of the branch, and the distance between sprinklers is 10 feet. The minimum pressure to any sprinkler is 10 psig. The discharge constant for the sprinklers is 5.6. What is the volume of water discharged from each of the three open sprinklers? Do not consider velocity pressure. (Velocity pressure is considered in Ex. 9.)

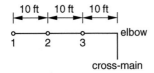

(Traditional U.S. Solution)

Follow the procedure in Sec. 76. Disregard steps involving velocity pressure.

     step 1: The normal (static) pressure at sprinkler 1 is 10 psig. From Eq. 2, the discharge from sprinkler 1 is

$$Q_1 = K\sqrt{p_{n,1}} = \left(5.6\,\frac{\text{gpm}}{\sqrt{\text{psig}}}\right)\sqrt{10\,\text{psig}}$$
$$= 17.71\,\text{gpm}$$

     step 2: The Hazen-Williams coefficient for steel pipe is $C = 120$. From App. B, the friction loss in 1-inch pipe with a flow rate of 17.7 gpm is approximately 0.10 psig per foot. The pressure loss in 10 feet of pipe is 1.0 psig.

Alternatively, Eq. 5 can be used. The inside diameter of a 1-inch pipe is 1.049 inches. The friction loss between sprinklers 1 and 2 is

$$p_{f,1\text{-}2} = \frac{(4.52)LQ^{1.85}}{C^{1.85}d^{4.87}}$$
$$= \frac{(4.52)(10\,\text{ft})(17.71\,\text{gpm})^{1.85}}{(120)^{1.85}(1.049\,\text{in})^{4.87}}$$
$$= (0.0051)(17.71\,\text{gpm})^{1.85}$$
$$= 1.04\,\text{psig}$$

     step 3: The total pressure at sprinkler 2 (see figure) is

$$p_{t,2} = p_{n,1} + p_{f,1\text{-}2} = 10\,\text{psig} + 1.04\,\text{psig}$$
$$= 11.04\,\text{psig}$$

     step 4: Skipped.

     step 5: The normal pressure to be used with sprinkler 2 is

$$p_{n,2} = p_{t,2} = 11.04\,\text{psig} \qquad [>10\,\text{psig, so OK}]$$

     step 6: The discharge from sprinkler 2 is

$$Q_2 = K\sqrt{p_{n,2}} = \left(5.6\,\frac{\text{gpm}}{\sqrt{\text{psig}}}\right)\sqrt{11.04\,\text{psig}}$$
$$= 18.61\,\text{gpm}$$

     step 7: The quantity flowing between sprinklers 2 and 3 is

$$Q_{2\text{-}3} = Q_1 + Q_2 = 17.71\,\text{gpm} + 18.61\,\text{gpm}$$
$$= 36.32\,\text{gpm}$$

The friction loss between sprinklers 2 and 3 is

$$p_{f,2\text{-}3} = (0.0051)(36.32\,\text{gpm})^{1.85}$$
$$= 3.93\,\text{psig}$$

     step 8: Skipped.

     step 9: The total pressure at sprinkler 3 is

$$p_{t,3} = p_{n,2} + p_{f,2\text{-}3}$$
$$= 11.04\,\text{psig} + 3.93\,\text{psig}$$
$$= 14.97\,\text{psig}$$

     step 10: Skipped.

     step 11: Skipped.

     step 12: Skipped.

     step 13: The normal pressure at sprinkler 3 is

$$p_{n,3} = p_{t,3} = 14.97\,\text{psig}$$

     step 14: The discharge from sprinkler 3 is

$$Q_3 = K\sqrt{p_{n,3}} = \left(5.6\,\frac{\text{gpm}}{\sqrt{\text{psig}}}\right)\sqrt{14.97\,\text{psig}}$$
$$= 21.67\,\text{gpm}$$

The standardized sprinkler systems worksheet (App. F), if used, would appear as follows (see next page).

Contract No. _____   Sheet No. ___1___ of ___1___

Name and Location _Example 66.8_ _____

nozzle type and location	flow gpm (l/min)	pipe size in	fitting and devices	pipe equivalent length	friction loss psi/ft (bars/m)	required pressure psi (bars)	normal pressure	K = 5.6 notes
1	q 17.71 Q 17.71		length 10 fittings 0 total 10		C = 120	p_t 10 p_t p_f 1.04 p_v p_e 0 p_n		$q = 5.6\sqrt{10} = 17.71$ steps 1–2
2	q 18.61 Q 36.32		length 10 fittings 0 total 10			p_t 11.04 p_t p_f 3.93 p_v p_e 0 p_n		$q = 5.6\sqrt{11.04} = 18.61$ steps 3–7
3	q 21.67 Q 57.99		length 10 fittings 0 total 10			p_t 14.97 p_t p_f p_v p_e p_n		$q = 5.6\sqrt{14.97} = 2.67$ steps 8–14

Example 9

Repeat Ex. 8 considering velocity pressure.

(Traditional U.S. Solution)

Follow the procedure in Sec. 76.

  step 1: The normal (static) pressure at sprinkler 1 is 10 psig. From Eq. 2, the discharge from sprinkler 1 is

$$Q_1 = K\sqrt{p_{n,1}} = \left(5.6\,\frac{\text{gpm}}{\sqrt{\text{psig}}}\right)\sqrt{10\,\text{psig}}$$

$$= 17.71\,\text{gpm}$$

  step 2: The Hazen-Williams coefficient for steel pipe is $C = 120$. From App. B, the friction loss in 1-inch pipe with a flow rate of 17.7 gpm is approximately 0.10 psig per foot. The pressure loss in 10 feet of pipe is 1.0 psig.

Alternatively, Eq. 5 can be used. The inside diameter of a 1-inch pipe is 1.049 inches. The friction loss between sprinklers 1 and 2 is

$$p_{f,1\text{-}2} = \frac{(4.52)LQ^{1.85}}{C^{1.85}d^{4.87}}$$

$$= \frac{(4.52)(10\,\text{ft})(17.71\,\text{gpm})^{1.85}}{(120)^{1.85}(1.049\,\text{in})^{4.87}}$$

$$= (0.0051)(17.71\,\text{gpm})^{1.85}$$

$$= 1.04\,\text{psig}$$

  step 3: The total pressure at sprinkler 2 (see figure) is

$$p_{t,2} = p_{n,1} + p_{f,1\text{-}2} = 10\,\text{psig} + 1.04\,\text{psig}$$

$$= 11.04\,\text{psig}$$

  step 4: The cross-sectional area of a 1-inch diameter schedule-40 steel pipe is 0.0060 ft^2. The velocity in the pipe between sprinklers 1 and 2 is

$$v = \frac{\dot{V}}{A} = \frac{(17.7\,\text{gpm})\left(\dfrac{0.002228\,\text{ft}^3}{\text{sec-gpm}}\right)}{0.0060\,\text{ft}^2}$$

$$= (17.7\,\text{gpm})\left(0.3713\,\frac{\text{ft}}{\text{sec-gpm}}\right)$$

$$= 6.57\,\text{ft/sec}$$

From Eq. 6, the velocity pressure of the flow between sprinklers 1 and 2 is

$$p_{v,1\text{-}2} = 0.433\,\frac{v^2}{2g}$$

$$= \frac{\left(0.433\,\dfrac{\text{psi}}{\text{ft}}\right)\left(6.57\,\dfrac{\text{ft}}{\text{sec}}\right)^2}{(2)\left(32.2\,\dfrac{\text{ft}}{\text{sec}^2}\right)}$$

$$= (6.724 \times 10^{-3})\left(6.57\,\frac{\text{ft}}{\text{sec}}\right)^2$$

$$= 0.29\,\text{psig}$$

step 5: The normal pressure to be used with sprinkler 2 is

$$p_{n,2} = p_{t,2} - p_{v,1\text{-}2} = 11.04 \text{ psig} - 0.29 \text{ psig}$$
$$= 10.75 \text{ psig} \quad [>10 \text{ psig, so OK}]$$

step 6: The discharge from sprinkler 2 is

$$Q_2 = K\sqrt{p_{n,2}} = \left(5.6 \frac{\text{gpm}}{\sqrt{\text{psig}}}\right)\sqrt{10.75 \text{ psig}}$$
$$= 18.36 \text{ gpm}$$

step 7: The quantity flowing between sprinklers 2 and 3 is

$$Q_{2\text{-}3} = Q_1 + Q_2 = 17.71 \text{ gpm} + 18.36 \text{ gpm}$$
$$= 36.07 \text{ gpm}$$

The friction loss between sprinklers 2 and 3 is

$$p_{f,2\text{-}3} = (0.0051)(36.07 \text{ gpm})^{1.85}$$
$$= 3.88 \text{ psig}$$

step 8: The velocity between sprinklers 2 and 3 is

$$v = (36.07 \text{ gpm})\left(0.3713 \frac{\text{ft}}{\text{sec-gpm}}\right)$$
$$= 13.4 \text{ ft/sec}$$

The velocity pressure between sprinklers 2 and 3 is

$$p_{v,2\text{-}3} = (6.724 \times 10^{-3})\left(13.4 \frac{\text{ft}}{\text{sec}}\right)^2$$
$$= 1.21 \text{ psig}$$

step 9: The total pressure at sprinkler 3 is

$$p_{t,3} = p_{n,2} + p_{f,2\text{-}3} + p_{v,2\text{-}3}$$
$$= 10.75 \text{ psig} + 3.88 \text{ psig} + 1.21 \text{ psig}$$
$$= 15.84 \text{ psig}$$

step 10: Estimate the discharge from sprinkler 3 as 20 gpm. The normal pressure that would cause this discharge is not calculated.

step 11: The estimated flow rate from the cross-main elbow to sprinkler 3 is

$$Q_{3,\text{elbow}} = Q_{2\text{-}3} + Q_{3,\text{estimated}}$$
$$= 36.07 \text{ gpm} + 20 \text{ gpm} = 56.07 \text{ gpm}$$

step 12: The estimated velocity in the pipe from the cross-main elbow to sprinkler 3 is

$$v_{\text{elbow-3, estimated}} = (56.07 \text{ gpm})\left(0.3713 \frac{\text{ft}}{\text{sec-gpm}}\right)$$
$$= 20.82 \text{ ft/sec}$$

The estimated velocity pressure between the cross-main elbow and sprinkler 3 is

$$p_{v,\text{elbow-3,estimated}} = (6.724 \times 10^{-3})\left(20.82 \frac{\text{ft}}{\text{sec}}\right)^2$$
$$= 2.91 \text{ psig}$$

step 13: The normal pressure at sprinkler 3 is

$$p_{n,3} = p_{t,3} - p_{v,\text{elbow-3,estimated}}$$
$$= 15.84 \text{ psig} - 2.91 \text{ psig} = 12.93 \text{ psig}$$

step 14: The corrected discharge from sprinkler 3 is

$$Q_{3,\text{corrected}} = K\sqrt{p_{n,3}} = \left(5.6 \frac{\text{gpm}}{\sqrt{\text{psig}}}\right)\sqrt{12.93 \text{ psig}}$$
$$= 20.13 \text{ gpm}$$

step 15: The discharge from sprinkler 3 was assumed in step 10 to be 20 gpm. It was calculated as 20.13 gpm in step 14. Assuming this is sufficiently close, 20.13 gpm would be used in subsequent steps.

The standardized sprinkler systems worksheet (App. F), if used, would appear as follows (see next page).

77. Hydraulic Design Procedure:
Pressure at Cross-Mains

The total pressure at the cross-main connection to a branch line is the normal pressure at the nearest open sprinkler plus the friction loss and the velocity pressure in the pipe between the sprinkler and the cross-main connection. Thus, the connection is assumed to be an end outlet (see Sec. 73) and the velocity downstream of the cross-main connection is used to calculate the total pressure. Minor losses, such as tee and nipple losses, must be included in the friction loss.

The pressure at each subsequent upstream cross-main-to-branch connection is calculated similarly to that for branch lines, except that the velocity head (already included from the most-remote cross-main connection) is assumed to be unchanged. The normal pressure at each

Contract No. _____   Sheet No. ____1____ of ____1____

Name and Location _Example 66.9_____

nozzle type and location	flow gpm (l/min)	pipe size in	fitting and devices		pipe equivalent length	friction loss psi/ft (bars/m)	required pressure psi (bars)		normal pressure	K = 5.6 notes
1	q 17.71 Q 17.71		length fittings total	10 0 10		C = 120	P_t 10 P_f 1.04 P_e 0	P_t P_v P_n		$q = 5.6\sqrt{10} = 17.71$ steps 1–2
2	q 18.36 Q 36.07		length fittings total	10 0 10			P_t 11.04 P_f 3.88 P_e 0	P_t 11.04 P_v 0.29 P_n 10.75		$q = 5.6\sqrt{10.75} = 18.36$ steps 3–7
3	q 20.13 Q 56.20		length fittings total	10 0 10			P_t 15.84 P_f P_e	P_t 15.84 P_v 2.91 P_n 12.93		$P_t = 10.75 + 3.88 + 1.21$ $= 15.84$ $q = 5.6\sqrt{12.93} = 20.13$ steps 8–14

successive cross-main-to-branch line connection is assumed to be the normal pressure of the last cross-main connection plus the friction pressure loss between the two branch connections.[26]

The analysis procedure works back up the cross-main from branch to subsequent upstream branch. While the normal pressure is known, the flow through each subsequent branch (not the hydraulically most-remote) is not.

If the subsequent branch is identical to a previous branch, the subsequent branch can be treated as an orifice with the orifice constant, K, calculated from the flow rate and normal pressure at the cross-main for the previous branch. All identical branches will have the same orifice constant. The discharge for the subsequent branch is simply calculated from Eq. 2.

If the subsequent branch is different from all previous branches, the new branch flow is initially calculated based on any conveniently assumed pressure at the end sprinkler in that branch, working up the branch back to the cross-main. The final calculated normal pressure will not coincide with the normal pressure calculated by working up the cross-main. This means that the calculated flow will also be incorrect.

Pressures at hydraulic junction points must balance within 0.5 psi (3.5 kPa; 0.035 bars) (NFPA 13, Sec. 6-4.3.1). Pressure differences and their corresponding

flows greater than this tolerance must be corrected. Orifice plates and sprinklers with mixed orifice sizes generally cannot be used to balance the system, although there are exceptions for small rooms (NFPA 13, Sec. 6-4.4.6).

The corrected flow quantity in the branch is calculated from Eq. 8.

$$\frac{Q_{\text{corrected}}}{Q_{\text{calculated}}} = \sqrt{\frac{p_{\text{cross-main}}}{p_{\text{calculated}}}} \qquad 8$$

When two branches on opposite sides of a cross-main have the same configuration, the flow calculated from one can be doubled. When the two opposite branches are different in configuration, the hydraulically shorter flow will have to be adjusted by using Eq. 8. The effect of this adjustment is to increase the flow in the shorter branch.

The corrected flow quantity is added to the cumulative flow in the cross-main. Then, the next upstream branch is handled similarly until all branches in the design area (see Sec. 63) are calculated. Once the flows from the design area have been calculated, the flow in the cross-main is assumed constant all the way back to the supply valve. The distance from the design area to the supply valve is used to calculate the pressure friction loss based on the discharge from all sprinklers in the design area. However, branches outside the design area do not influence the flow rate in the cross-main.

[26]This is normally a valid assumption, since the velocities in the cross-mains will be low. However, if necessary, a rigorous calculation based on actual velocities can be performed. This is seldom necessary.

Example 10

The sprinkler branch from Ex. 9 is fed by a $1\frac{1}{2}$-inch (nominal) schedule-40 steel pipe cross-main. What are the total and normal pressures in the cross-main at the entrance to the elbow?

branch A
○———┐ elbow
3   │
       │

(Solution)

From Ex. 9, the flow rate between the elbow and sprinkler 3 is

$$Q_{\text{elbow-3}} = 36.07 \text{ gpm} + 20.13 \text{ gpm} = 56.20 \text{ gpm}$$

The velocity between sprinkler 3 and the elbow is

$$v = (56.20 \text{ gpm})\left(0.3713 \frac{\text{ft}}{\text{sec-gpm}}\right)$$
$$= 20.87 \text{ ft/sec}$$

The velocity pressure between sprinkler 3 and the elbow is

$$p_{v,\text{elbow-3}} = (6.724 \times 10^{-3})\left(20.87 \frac{\text{ft}}{\text{sec}}\right)^2$$
$$= 2.93 \text{ psig}$$

Since the flow is turned 90°, the minor loss of the elbow (based on the smaller diameter) must be included with the friction loss. The equivalent length of a 1-inch elbow is given in App. D as 2 feet. The total friction loss between sprinkler 3 and the elbow is

$$p_{f,\text{elbow-3}} = \frac{(4.52)LQ^{1.85}}{C^{1.85}d^{4.87}}$$
$$= \frac{(4.52)(10 \text{ ft} + 2 \text{ ft})(56.20 \text{ gpm})^{1.85}}{(120)^{1.85}(1.049 \text{ in})^{4.87}}$$
$$= 10.56 \text{ psig}$$

The total pressure at the end of the cross-main at the entrance to the elbow at branch A is

$$p_{t,\text{A}} = p_{n,3} + p_{v,\text{elbow-3}} + p_{f,\text{elbow-3}}$$
$$= 12.93 \text{ psig} + 2.93 \text{ psig} + 10.56 \text{ psig}$$
$$= 26.42 \text{ psig}$$

The normal pressure at the end of the cross-main at the entrance to the elbow at branch A excludes the velocity pressure.

$$p_{n,\text{A}} = p_{n,3} + p_{f,\text{elbow-3}}$$
$$= 12.93 \text{ psig} + 10.56 \text{ psig}$$
$$= 23.49 \text{ psig}$$

Example 11

The $1\frac{1}{2}$-in (nominal) cross-main evaluated in Exs. 9 and 10 contains two additional opposing branches. Both branches use 1-inch (nominal) schedule-40 steel pipe. The two opposing branches have different configurations. The branch flow rates and normal pressures at the cross connection based on an assumed minimum pressure of $p_{\text{minimum}} = 10$ psig are given. The distance along the cross-main between branches A and B/C is 10 ft. Branch A, evaluated in Ex. 10, is the hydraulically most-remote. What is the total flow rate in the cross-main?

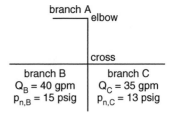

(Solution)

The diameter of $1\frac{1}{2}$-inch-diameter pipe is 1.610 inches.

The minor loss of the elbow has already been included. The cross is a straight-through connection, and its minor loss is disregarded. The friction pressure loss between the elbow and the cross is

$$p_{f,\text{elbow-cross}} = \frac{(4.52)LQ^{1.85}}{C^{1.85}d^{4.87}}$$
$$= \frac{(4.52)(10 \text{ ft})(56.20 \text{ gpm})^{1.85}}{(120)^{1.85}(1.610 \text{ in})^{4.87}}$$
$$= 1.09 \text{ psig}$$

The normal pressure at the cross is

$$p_{n,\text{cross}} = p_{n,\text{elbow}} + p_{f,\text{elbow-cross}}$$
$$= 23.49 + 1.09 \text{ psig} = 24.58 \text{ psig}$$

The calculation for the normal pressure for branches B and C was not 24.58 psig. Therefore, the residual pressure at the ends of branches B and C will not be 10 psig—it will be higher. The flow rates are also incorrect. Equation 8 is used to correct the flows.

$$Q_{\text{B,corrected}} = (Q_{\text{B,calculated}})\sqrt{\frac{p_{\text{cross-main}}}{p_{\text{calculated}}}}$$
$$= (40 \text{ gpm})\sqrt{\frac{24.58 \text{ psig}}{15 \text{ psig}}} = 51.2 \text{ gpm}$$
$$Q_{\text{C,corrected}} = (35 \text{ gpm})\sqrt{\frac{24.58 \text{ psig}}{13 \text{ psig}}} = 48.1 \text{ gpm}$$

When branches A, B, and C are all open, the total flow into the cross main will be

$$Q_{\text{total}} = 56.2 \text{ gpm} + 51.2 \text{ gpm} + 48.1 \text{ gpm} = 155.5 \text{ gpm}$$

78. Hydraulic Design Procedure: Pressure in Risers

The total pressure at the top of a riser is calculated by adding the normal pressure at the nearest (downstream) flowing branch, the total friction loss between the branch and the top of the riser (including the minor loss for the riser-to-cross-main fitting), and the velocity pressure in the cross-main at the riser connection. Thus, the total pressure is based on the velocity downstream of the riser-to-cross-main connection.

The pressure at the bottom of a riser is the pressure at the top of the riser plus the friction loss in the riser plus the elevation (i.e., static) pressure corresponding to the change in elevation. (Each foot of height corresponds to approximately 0.434 psi.)

Example 12

The $1\frac{1}{2}$-inch-diameter cross-main of Exs. 9 and 10 continues 10 feet and then connects to a $1\frac{1}{2}$-inch-diameter, 12-foot-high riser. The riser is fed by a 3-inch feed main at the lower level. The riser-to-cross-main connection at the top of the riser is through a standard 90° elbow. The connection to the feed main at the base of the riser is through a tee. What are the normal pressures at the top and base of the riser?

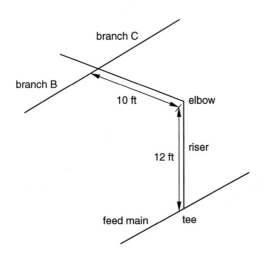

(Solution)

The equivalent length of a standard $1\frac{1}{2}$-inch 90° elbow is given in App. D as 4 feet. The friction loss between branches B and C and the top of the riser is

$$p_{f,\text{cross-top}} = \frac{(4.52)LQ^{1.85}}{C^{1.85}d^{4.87}}$$
$$= \frac{(4.52)(10 \text{ ft} + 4 \text{ ft})(155.5 \text{ gpm})^{1.85}}{(120)^{1.85}(1.610 \text{ in})^{4.87}}$$
$$= 10.05 \text{ psig}$$

The normal pressure at the top of the riser is

$$p_{n,\text{top}} = p_{n,\text{cross}} + p_{f,\text{cross-top}}$$
$$= 24.58 \text{ psig} + 10.05 \text{ psig} = 34.63 \text{ psig}$$

The friction loss in the 12 feet of riser is

$$p_{f,\text{riser}} = \frac{(4.52)LQ^{1.85}}{C^{1.85}d^{4.87}}$$
$$= \frac{(4.52)(12 \text{ ft})(155.5 \text{ gpm})^{1.85}}{(120)^{1.85}(1.610 \text{ in})^{4.87}}$$
$$= 8.62 \text{ psig}$$

The term *base of the riser* is somewhat ambiguous and probably means the feed main. Thus the pressure at the base of the riser is the required feed main pressure. From App. D, the equivalent length of a $1\frac{1}{2}$-in tee is 8 feet. The friction loss in the tee is

$$p_{f,\text{tee}} = \frac{(4.52)LQ^{1.85}}{C^{1.85}d^{4.87}}$$
$$= \frac{(4.52)(8 \text{ ft})(155.5 \text{ gpm})^{1.85}}{(120)^{1.85}(1.610 \text{ in})^{4.87}}$$
$$= 5.74 \text{ psig}$$

In addition, the elevation pressure head from 12 feet of water is

$$p_{\text{elevation}} = \gamma h$$
$$= \frac{\left(62.4 \dfrac{\text{lbf}}{\text{ft}^3}\right)(12 \text{ ft})}{144 \dfrac{\text{in}^2}{\text{ft}^2}}$$
$$= 5.20 \text{ psig}$$

The pressure at the base of the riser is

$$p_{n,\text{base}} = p_{n,\text{top}} + p_{f,\text{riser}} + p_{f,\text{tee}} + p_{\text{elevation}}$$
$$= 34.63 \text{ psig} + 8.62 \text{ psig} + 5.74 \text{ psig} + 5.20 \text{ psig}$$
$$= 54.19 \text{ psig}$$

79. Allowances for Hose Streams

The municipal water supply connection to the fire protection system must have the capacity to support hose lines and hydrants that are open at the same time as the sprinkler system.

Table 16 contains hose stream allowances for hydraulically-designed systems. Each hose stream is allotted 50 gpm (3.2 ℓ/s; 190 ℓ/min) (NFPA 13, Sec. 5-2.3.-1.3(d)), and the first two hose stations are inside the building and operate at the pressure of the sprinkler system. The remainder of the hose stream allowance (i.e., the outside hose allowance) is added to the sprinkler and inside hose requirement at the connection with the municipal water supply, or is given to the yard hydrant (depending on which is the closer to the system riser) (NFPA 13, Sec. 5-2.3.1.4(f)).

Inside and outside hose stream allowances are not necessary when gravity, pressure tanks, or pumps supply only sprinklers (NFPA 13, Sec. 5-2.3.1.1, Exceptions 2 and 3; Sec. 5-2.3.1.3(h)).

Table 16
Hose Stream Allowances vs. Occupancy Type
(hydraulically calculated systems)

occupancy classification	inside hose, gpm (ℓ/s)	combined (inside and outside hose), gpm (ℓ/s)	required duration, min
light	0, 50, 100	100	30
	(0, 3, 6)	(6)	
ordinary	0, 50, 100	250	60–90
groups 1 & 2	(0, 3, 6)	(16)	
extra hazard	0, 50, 100	500	90–120
groups 1 & 2	(0, 3, 6)	(32)	

Multiply psi by 6.895 to obtain kPa.
Multiply psi by 0.0689 to obtain bars.
Multiply gpm by 0.0631 to obtain ℓ/s.
Multiply gpm by 3.785 to obtain ℓ/min.

Reprinted with permission from NFPA 13, *Standard for the Installation of Sprinkler Systems*, copyright © 1991, National Fire Protection Association, Quincy, MA 02269. This reprinted material is not the complete and official position of the National Fire Protection Association on the referenced subject, which is represented only by the standard in its entirety.

The lower duration values in Table 16 are applicable when water flow is monitored by a central or remote station (NFPA 13, Sec. 5-2.3.1.3 (g)).

80. Nozzles for Hand Line Hoses

Hose *nozzles*, more appropriately referred to in fire-fighting terminology as *assemblies, appliances,* or *devices,* may consist of barrels, reducers, and various tips. The pattern and volume of water from a hose depends on the nozzle used. (Physically larger nozzles do not always flow greater quantities, however.) Nozzles on $1\frac{1}{2}$-inch lines generally discharge at the rate of 50–125 gpm (3.2–7.9 ℓ/s), while $2\frac{1}{2}$ line nozzles are most efficient in the 200–300 gpm (13–19 ℓ/s) range.

Conventional nozzles are of the solid-stream nozzle variety. *Solid-stream nozzles* are identified by the inside diameter of their tips and the hose size they are intended for. The standard solid-stream nozzle for a $2\frac{1}{2}$-inch hand line has a $1\frac{1}{8}$-inch tip and flows approximately 250 gpm (16 ℓ/s; 950 ℓ/min). The minimum recommended tip size for hand lines is 1 inch. Also, $1\frac{1}{4}$-inch tips are available. Solid-stream nozzles may not be approved for all uses.

Spray nozzles are identified by the rates of discharge at standard pressures of 50 and 100 psig (345 and 690 kPa; 3.45 and 6.9 bars). There are three general types of spray nozzles:

1. open nozzles with fixed spray patterns

2. adjustable nozzles capable of discharge patterns from solid stream through narrow and wide spray to shutoff

3. combination nozzles where the solid stream, spray, or shutoff is selected by control valve

A pressure of 25 psi (172 kPa; 1.72 bars) is typical for civilian hand line and standpipe hose use. Because of the hose reaction, higher pressures in the 45–50 psig (310–345 kPa; 3.1–3.45 bars) range are intended for fire department use.

The *discharge coefficient* for a nozzle is the ratio of the actual to theoretical volumetric discharge. Typical values of the discharge coefficient are: standard square-edged orifice, 0.61; standard $\frac{1}{2}$-in sprinkler, 0.75; short fire-fighting shut-off nozzle, 0.90 short playpipe nozzle, 0.96; general smooth-bore nozzle, 0.96-0.98; Underwriters' playpipe, and 0.97; and deluge (monitor) nozzle, 0.99.

81. Friction Losses in Hoses

The Hazen-Williams equation (Sec. 71) can be used to calculate friction losses in rubber-lined fire hose.

82. Inspection and Maintenance of Fire Sprinkler Systems

Repairs are rarely needed in sprinkler systems because the system components are not subject to wear or operating stress. The main exceptions are rubber valve seats, air compressors, and fire pumps. Failures of a sprinkler system can usually be traced to closed valves, bent or obstructed sprinklers, frozen piping, inoperative fire pumps, and deteriorated or impaired water supplies.

Repair and maintenance of a fire sprinkler system (known as an *impairment*) must be carefully planned and coordinated to minimize the time the area is unprotected.

Most sprinkler failures can be avoided by implementing a regular inspection procedure. While the actual inspection elements will vary depending on the installation, Table 17 can be used as the basis for a more specific plan.

Table 17
General Sprinkler System Inspection Schedule

Daily:

- control valves are open
- external Siamese connections are unblocked and undamaged
- water pressure, water levels, and air pressure are in normal range
- fire hose racks and reels are unobstructed
- hoses are present and properly stowed
- aisles are clear
- nothing is hanging from sprinkler piping
- no conditions are present causing piping to freeze

Weekly:

- inspect sealed control valves
- check air and water pressures
- test operation of fire pumps
- check engine fuel, oil, coolant, and battery water levels
- check specific gravity of battery water
- test pump relief valve
- test pump discharge pressure
- test emergency power transfer

Monthly:

- inspect fire department connections
- inspect locked control valves
- inspect tamper switches
- test supervisory devices
- drain and service air compressor tanks and moisture traps

Quarterly:

- test waterflow alarms
- flow test main drain
- inspect pump priming water level

Semiannually:

- test preaction and deluge detection systems
- test quick-opening devices
- test detections devices (e.g., smoke detectors)

Annually:

- perform preventive maintenance on control valves
- test open sprinklers
- test specific gravity of antifreeze solutions in wet systems
- test low-point drains
- trip test dry pipe valves
- test fire pump at full discharge
- open low-point drains to remove condensate
- clean and grease battery posts and cable clamps

Other:

- flush piping every 5 years
- test high-temperature sprinklers every 5 years
- test residential sprinklers every 20 years
- replace standard sprinklers every 50 years

PART 5:
Explosion Protection Systems

83. Introduction to Explosions

An *explosion* requires a combustible material, air or other oxidant, and a source of ignition. Three categories of combustible substances are prone to explosions: gases and vapors, dusts, and liquid mists and aerosols. Other less-common types of explosions are pressure-release explosions involving pressurized containers and decomposition explosions involving instantaneous decomposition of certain endothermic compounds such as explosives and ammunition. (Pressure vessel explosions often do not cause a fire.)

Dust explosions are not as common as vapor-related explosions. However, certain industries and operations seem to be more prone to dust explosions than others. These include grain elevators, flour mills, candy factories, paint manufacturing operations, and metal powder operations.

Experience has shown that explosions in one process vessel will propagate into adjacent vessels through connecting piping. When the flame front velocity in an explosion is less than the speed of sound at the appropriate pressure, the explosion is called a *deflagration*. When the flame speed is greater than the speed of sound, the explosion is more violent and is called a *detonation*. Events involving flammable gases can be either explosions or deflagrations. The majority of industrial dust explosions are deflagrations.

84. Secondary Explosions

Secondary explosions occur a noticeable time after the initial explosion, with their ignition source being the initial explosion. Even when venting techniques have successfully operated, feed lines to an explosion point may continue to supply fresh combustible material to a post-explosion fire. Also, smoldering material from the first explosion may be carried downstream through the process piping or ducting. Both of the scenarios can result in secondary explosions. Detection and isolation devices are necessary to prevent secondary explosions.

85. Explosive Conditions

Most explosive conditions result from abnormal operations. These include:

- transient phases of chemical processes (e.g., when solvents are piped in and out of vessels)
- pneumatic addition of solid materials
- pneumatic addition of solid materials to vessels containing flammable solvents
- manual inspection, when vessels are opened to the atmosphere
- vent and leak points

Explosive atmospheres do not have to be an intrinsic part of a manufacturing process for hazard to be present. For example, high-pressure leaks (such as from hydraulic oil lines) and the agitation of low-viscosity liquids can generate explosive atmospheres. Fugitive emissions (i.e., releases of flammable vapors from such sources as leaks around pump seals, valve packing, flange gaskets, compressor seals, and process drains) can also constitute explosion hazards.

86. Flash Point

The *flash point* of a flammable mixture is the lowest temperature at which a substance in an open vessel will ignite in a momentary flash when flame is brought near it. To be self-sustaining, the temperature must be slightly higher than the flash point. The minimum ignition temperature in process vessels, where the conditions may be significantly different, will vary widely.

It is the vapors of flammable liquids, not the liquids themselves, that burn and explode. The flash point is the minimum temperature at which the liquid gives off enough vapor to form an ignitable mixture with air near the surface of the liquid.

Appendix E lists the flash points of common liquids.

87. Combustible and Flammable Liquids

A distinction is made between flammable liquids and combustible liquids. *Flammable liquids* have flash points below 100°F (38°C) and vapor pressure not exceeding 40 psia (2068 mm Hg) at 100°F (38°C). Flammable liquids are also known as Class I liquids, with a further subclassification based on flash point and boiling point. Class IA liquids have flash points below 73°F (23°C) and boiling points below 100°F (38°C). Class IB liquids have flash points below 73°F (23°C) and boiling points above 100°F (38°C). Class IC liquids have flash points at or above 73°F (23°C) and below 100°F (38°C).

Combustible liquids have flash points above 100°F (38°C). A further classification of combustible liquids is made on the basis of flash point. Class II liquids have flash points in the range of 100–140°F (38–60°C). Class IIIA liquids have flash points in the range of 140–200°F (60–93°C). Class IIIB liquids have flash points above 200°F (93°C).

88. Fire Point

The *fire point* is generally a few degrees above the flash point. Vapors will ignite at the flash point temperature, but the fire will not be sustaining because the liquid portion is not generating vapors fast enough.

The difference between the flash and fire points is small. Therefore, the fire point is not of major importance in designing protective systems for hazardous liquids.

89. Vapor Density: Vapors and Dusts

Vapor density, as the term is used in the hazardous vapor field, usually refers to the specific gravity of the vapor as measured with respect to air.[27] If unconstrained, vapors with specific gravities less than 1.0 (e.g., natural gas) tend to rise and remove themselves from possible ignition sources at lower levels.

When referring to combustible dusts, the density is given in terms of weight or mass per cubic foot of air. Typical units are oz/ft^3 (g/m^3).

90. Explosive Limits

Most substances do not become explosive until mixed with an oxidizer. Air normally provides the oxygen

required for combustion. For each hazardous gas, there is a range of volumetric fuel:air ratios that constitute a risk. For common gases, the ratio is in the 1–15% range by volume. For any specific substance, the low limit is known as the *lower explosive limit* (LEL); the high limit is known as the *upper explosive limit* (UEL).

The range between the LEL and UEL is generally narrow. For gasoline at normal temperatures and pressures, for example, the LEL and UEL are approximately 1.3 and 6% by volume, respectively. For natural gas (methane), the range is between 5 and 15%. Notable exceptions are acetylene (2.5 and 100%) and carbon disulfide (1 and 60%). Other values are given in App. E.

Explosive limits for a dust are expressed as a density, not a volumetric percentage. The particle size is an important factor in determining the explosive limit concentrations. LEL values for dusts are more easily determined; UEL values are more difficult to quantify. Typical values are in the vicinity of 0.02 oz/ft^3 (20 g/m^3). As a general rule, when visibility in a dusty atmosphere has been reduced to a few meters, the LEL has been exceeded.

91. Ignition Temperature

The temperature necessary to produce ignition in a small portion of a vapor is called the *ignition temperature (self-ignition, autoignition,* and *autogeneous ignition temperature)*. A small spark may be sufficient to raise the temperature of a small portion of the vapor to the ignition temperature. The ignition temperature is usually several hundred degrees above the flash point.

92. Factors Affecting Dust Explosions

Some dusts explode, and others do not. Numerical characteristics describing the explosive potential of each dust include the *maximum explosion pressure* (p_{max}) and the *maximum rate of pressure rise (dp/dt)*. Pressures as high as 150 psi (1035 kN/m^2) and rates of rise of 20,000 psi/sec (140 000 kN/m$^2 \cdot$ s) have been noted for dust explosions. Other information that is useful includes the minimum ignition temperature, the minimum dust concentration, the minimum ignition energy, and the minimum required oxygen concentration.

The *minimum ignition energy* (MIE) is the minimum spark energy that will ignite and propagate a flame through the material. The smaller the average particle size, the lower the MIE. As little as 1 mJ will be sufficient to ignite some dusts. Table 18 lists typical required ignition energies. These values can be compared to the energies in Table 21.

---

[27]The vapor density of air is taken as 1.0. A vapor with a density of 1.5 will be one and a half times as heavy as air.

Table 18
Typical Minimum Ignition Energies

material	MIE (mJ)
oxygenated vapors	0.01 or less
hydrocarbon solvents	0.1
metallic dusts	1
fine aerosols	1
fine powders	5
course powders	60

In addition to the intrinsic flammability of the dust, other factors affecting whether a dust explodes includes the size of the source of ignition, the particle size distribution, the turbulence, moisture content, and purity of the dust.

93. Explosion Prevention Systems

Explosion prevention systems protect facilities and personnel. The two major protection techniques are venting and isolation. *Explosion vents* open instantaneously to dissipate explosion pressure. *Isolation* is used to arrest, divert, or extinguish flames in order to keep the explosion from spreading throughout the connecting pipes, ducts, and vessels. Isolation devices, primarily effective in deflagrations (see Sec. 83), are installed in piping and ducts between vessels.

Explosion prevention systems cannot completely eliminate explosion hazards. However, a combination of venting and isolation in conjunction with preventive measures will reduce the probability of damage and injury from an explosion to acceptably low levels.

94. Explosion Hazard Classification

Local, state, and federal authorities have some latitude in classifying hazards.[28] Dusts from high explosives, propellants, and oxidizers (which are not in Table 19) must be carefully considered in making the classification. However, there is no latitude in determining the level of protection afforded by a piece of equipment. Protection classifications such as "intrinsically safe" (Sec. 95) on equipment must be granted by a NRTL (see Sec. 17).

Table 19 summarizes the classes of equipment hazard occupancy that are presented in the NEC (National Electric Code Article 500). *Classification of Class I Hazardous Locations for Electrical Installations* (NFPA 497A) and *Classification of Class II Hazardous Locations for Electrical Installations* (NFPA 497B) are useful in classifying environments.

---

[28]Usually, however, this latitude results in a more strict classification of the hazard level.

Table 19
Equipment Hazard Classes

class	description
I	flammable liquids or gases that are handled, processed, transferred, and stored, or that may enter the air in ignitable concentrations
II	combustible dusts such as metal dusts (e.g., magnesium and aluminum), coal, charcoal, coke, wheat, flour, grain, and starch
III	easily ignitable fibers, shavings (e.g., sawdust, wood shavings, and textile fibers), and flyings that might be present, but are not likely to be in suspension in the air

The three atmospheric conditions are further divided into two divisions, depending on the presence of the hazardous substance (see Sec. 95 for more information on the divisions). For example, equipment approved for Division 1 occupancies is installed where the hazardous product is in the air during normal operation and processing. Examples of Division 1 occupancies are locations where hazardous material is handled in open containers, near fuel tank fills or vent ports, and in low-lying areas where heavier-than-air gases could accumulate (e.g., pump pits or enclosed spaces).

The NEC (which is the same as NFPA 70) gives general guidelines for classifying equipment. For actually designing equipment to operate in Division 1 and 2 environments, more specific guidelines are required. Such guidelines are typically developed, adopted, and prescribed by testing labs and insurers (e.g., UL and FM). For example, Division 1 motors are often squirrel-cage induction motors that are designed to UL or FM standards.

Table 20
Groups of Hazardous Commodities

group	hazards
A	acetylene
B	hydrogen and other gases of equivalent hazard
C	ethylene, ether, hydrazines
D	gasoline, alcohol, jet fuel, most flammable solvents, most flammable paints

Finally, specific hazards can be broken down into groups (as illustrated in Table 20) depending on their relative hazard. This relative hazard is based to a large extent on the range of fuel-air ratios that constitute an explosion hazard.

95. Explosion-Proof Equipment

Explosion-proof equipment generally refers to electrical devices and appliances approved for use in potentially explosive atmospheres. Approval is on an occupancy class-by-occupancy class basis.[29] Approval for use in one class does not constitute approval for another class. Thus, the term *approved for location* is more appropriate than explosion-proof.

In reality, it is not possible to guarantee that a piece of equipment will be totally safe. It is the object of the testing process, however, that the equipment is *single-fault tolerant*. That is, more than one thing must go wrong in order for the equipment to become a source of ignition. The occupancy division is used to specify the degree of fault tolerance in an environment.

In Division 1 environments (e.g., Class I, Division 1, or Class II, Division 1), the first fault would be with the equipment since the hazardous material is always present. While Division 1 equipment is sealed in a housing to keep the hazardous material from entering it, housing must also be heavy enough to keep the first fault (e.g., a spark or overheating element) from propagating beyond the housing and igniting the hazardous material outside. Since Division 1 housings are expected to contain and confine blasts occurring within their bodies, such housings are commonly referred to as being explosion-proof.

Equipment approved for Division 2 use (e.g., Class I, Division 2, or Class II, Division 2) is installed where the hazardous product is in the air only during abnormal operation. Examples of Division 2 uses are installations where the hazardous material is stored in closed containers, is present in the air only due to a leak, or is immediately adjacent to a Division 1 area.

Division 2 housings are also expected to keep equipment separated from the external housing. However, the housing is not expected to contain an explosion. The housing is expected only to isolate the equipment from the hazardous environment in normal operation. The first fault would be the presence of the hazardous material. A spark or overheated element within a Division 2 housing would be a second fault.

Some types of equipment do not need housings at all and are known as *intrinsically safe* items. Such items generally use only small amounts of energy. The energy available in intrinsically safe equipment is so low

that it would not be sufficient to ignite the hazardous atmosphere (NEC Article 500-1).

Another category is *purged and pressurized equipment*, which relies on a source of clean air or inert gas under pressure to prevent ignition. This method is particularly suitable for large areas such as control rooms. Built-in safeguards must shut down the equipment in the event of a loss of ventilation or pressurization.

There are three types of purging: X, Y, and Z. *X-purging* reduces the classification from Division 1 to nonhazardous, *Y-purging* reduces the classification from Division 1 to Division 2, and *Z-purging* reduces the classification from Division 2 to nonhazardous.

There are a few exceptions to the mandatory use of explosion-proof devices. One exception is where other ignition sources are present, such as welding shops that contain acetylene. Welding flames are an unavoidable ignition source, eliminating the need for protected electrical equipment. Other exceptions are made for flammable liquids that are kept below the flash point. *Pyphoric* (self-igniting) *materials* are also treated separately.

96. Sources of Ignition

To prevent explosions, likely sources of ignition must be identified and eliminated. As with fires, explosions can occur spontaneously in some bulk materials. However, most explosions can be traced to specific ignition sources.

Dryers, grinders, furnaces, kilns, ovens, and welding operations produce flames and smoldering particles that can ignite hazardous atmospheres. Friction in moving equipment (such as in mechanical handling equipment—e.g., conveyors) creates hot surfaces that can ignite a dust cloud.

Furthermore, the movement of air in dust collection ducts creates static electricity charges that can reach or exceed 50 mJ. Depending on the minimum ignition energy (see Table 18), static electricity may be all that is needed to ignite a hazardous atmosphere.

97. Static Electricity

Since static discharges do not contain enough energy to raise the bulk temperature high enough to start a fire, static electricity (i.e., incendiary electrostatic discharge) is not normally a problem with solvents and volatile liquids below their flash points. Fine mists, aerosols, and dusts, however, can be ignited by sparks of static electricity. For example, the flash point of kerosene in liquid form is approximately 100°F (38°C), although

---

[29]For the purposes of equipment classification, classes of flammable and combustible substances are defined in the NEC and presented in Sec. 94. For example, Class I occupancies are where flammable liquids, gases, and vapors exist.

kerosene mists can be ignited at 70°F (20°C) or more below the flash point.

The most common sources of electrostatic discharges are isolated conductors. These conductors acquire high-potential charges from friction, charge induction, or conduction. For this reason, all electrical components must be grounded.[30]

In addition to electrical components, all mechanical equipment, conveyors, air handling equipment, and dust collection ducts and components must be grounded. In particular, rotating machinery (such as centrifuges) must be grounded. *Epitropic filter fabric* (a fabric that has been impregnated with carbon and is highly conductive) should be used to disperse any static charges that build up in dust collectors.

All grounding must be to an effective grounding point outside of the equipment being safeguarded. For materials with minimum ignition energies of 100 mJ and below, the maximum resistance to ground should be approximately 10^8 ohms. The resistance can be checked with a standard resistance meter.

For a charge to develop, some part of the system must be electrically isolating. Many industrial powders, as well as some organic solvents (e.g., the aliphatic and aromatic compounds), are insulators. The increased use of plastic and plastic-lined pipes is a particular concern. Special consideration must be given to vessels and pipes that are lined with thin plastic layers.[31] Such configurations can effectively act as large capacitors, developing immense surface charge densities. The lightning-like sparks arising from these configurations are known as *propagating brush discharges*.

The human body can also act as an isolated conductor. Worse, people are mobile, creating a dangerous potential that can be carried into hazardous areas. Therefore, static electricity generated by individual workers must also be considered. Personnel involved in pouring, sieving, or transferring granular or powdered materials can themselves become charged with 10,000 volts or more. Workers in hazardous atmospheres should always wear conductive (i.e., antistatic) clothing and footwear.

---

[30] A ground is referred to in some codes as a *bond*. Thus being bonded is the same as being grounded.

[31] Pipes lined with polytetrafluoroethylene (PTFE) are examples of this.

Table 21
Typical Capacitances and Spark Energies

item	capacitance (in pF)	spark energy (in mJ at 20 kV)
small metal items (e.g., scoops, spoons)	10–20	2–4
small containers (volume < 0.5 m³)	10–100	2–20
large containers (volume > 0.5 m³)	50–300	10–60
human body	200–300	40–60
large equipment (e.g., chemical reactor vessels)	100–1000	20–200

Large industrial-sized ionizers can be used to neutralize highly charged atmospheres and equipment. However, the effectiveness of such devices is variable.

98. Inerting and Purging

Replacing all or part of atmospheric air with an inert gas can prevent an explosion by removing the oxidizer. This is known as *inerting* or *purging*. The goal is to reduce the oxygen content below the minimum level required for ignition. Due to the loss of the inert gas, however, this method is not suitable for open-circuit systems (i.e., those that bring in and discharge air).

99. Fire Protection Guidelines for Dust Hazards

Hose connections should be equipped with fine spray or fog nozzles. A solid or coarse water stream could throw more dust into suspension, resulting in additional fire or explosion damage. Where electrical equipment is in use, hoses should use special fog nozzles.

When metal dusts are the hazard, quantities of sand, talc, or other inert materials should be available to smother small fires. Water and portable (pressurized) extinguishers should not be used.

OSHA prescribes mandatory measures to avoid and minimize the effects of dust explosions. These include use of hoods, ducting, and dust-collection devices.

100. Explosion Suppression

The goal of explosion suppression is to stop an explosion that has already begun. All explosions are disruptive and are candidates for suppression. However, suppression is particularly attractive when toxic materials are being handled, since the distribution of toxic material creates a secondary hazard long after the explosion has occurred.

Explosion suppression depends on early detection of an explosion, usually within the first 10 milliseconds. Once detected, the control panel fires a small explosive squib in the suppression device. The shock wave from this explosion bursts a rupture valve, allowing a pressurized chemical suppressant to be injected into the duct or vessel. This suppressant absorbs the thermal energy produced by the flame, dilutes the fuel, displaces the oxygen, and impedes the combustion chemical reaction. Typical suppressants include water, halogenated hydrocarbons, ammonium phosphate, sodium bicarbonate, and sodium chloride. In recognition of their operating method, such systems are sometimes referred to as *chemical-responsive systems*.

Explosion suppression devices use rapid-acting valves. Detectors should be located at inlets and outlets of critical points, such as the inlet and outlet headers of dust collectors.

Residence time of the extinguishing agent must be evaluated. Without sufficient residence time, the agent may be blown through the system without completely extinguishing the reaction. This is particularly a concern with high volumetric flow rates.

101. Explosion Containment

The concept of protection by explosion containment is based on keeping the vessel (e.g., a mixing chamber, dust collector, or pipe) intact after the explosion. This requires the vessel to be able to withstand the maximum pressure generated during the explosion.

102. Explosion Relief

Explosion relief (explosion venting) is the planned venting of high pressures in a vessel, pipe, or duct. As the pressure increases, a relief vent, panel, or valve opens to allow the expanding gases to escape.

In the case of large vessels such as dust collectors, two types of explosion vents are available: explosion-relief doors (including pop-out panels) and bursting-panel relief vents. *Explosion-relief doors* are lightweight and retained by springs, magnets, or gravity. After an explosion, the door is merely returned to its original position. *Bursting-panel relief vents*, however, must be replaced.

When relying on explosion relief to minimize damage from explosions, attention must be given to secondary effects. The most obvious consideration is the area to where the explosion will be vented. Venting an explosion to another hazardous atmosphere or to an occupied area is unacceptable. Ideally, the entire vessel should be located outside of an occupied building. If the vessel must be inside, venting to the outside of the building is the preferred solution. This will require placing the vessel close to the exterior walls. Also, care must be taken to avoid having whole, or pieces of, relief doors and panels fly through air. Large pieces should be restrained by short lengths of wire rope.

103. Preventing Explosion Propagation

In addition to chemical extinguishment, the two primary methods of preventing the propagation of explosions are mechanical and passive systems.

Mechanical-responsive systems provide a mechanical barrier to flame. The most common mechanical-responsive system incorporates an explosion-isolation valve. Key components are the pressure detector located inside the vessel, an automatic control panel, and a fast-acting gate valve. When the explosion has been detected, the control panel fires a small explosive squib. The shock wave from the squib opens a rupture disk, allowing a rapid release of nitrogen to push down a piston, which in turn, closes a sliding-gate (knife) valve. The gate is the physical barrier that halts the flame front.

Explosion-isolation valves are available in a wide range of sizes, including 2–36 inches (5–91 cm) in diameter. Depending on size and manufacturer, typical response times are from 20–100 milliseconds. This is normally adequate for deflagrations but not for detonations (which are faster and more violent).

Passive-mechanical isolation systems do not require detection or control. The typical passive system is the flame-front diverter (backflash interrupter). Such devices place an acute bend in the piping. Flow at the normal velocity is able to navigate the bend, which a high-velocity explosion is not. The release dome or door may also include a trip switch that automatically shuts other valves or halts the processing.

Passive systems must not be installed in occupied areas. Their effectiveness is also limited to the processing of nonhazardous materials. If the product being produced is too hazardous for atmospheric release, explosion-isolation and chemical suppressants should be considered.

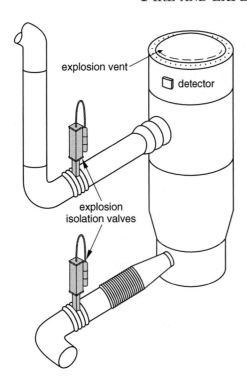

Figure 10 Mechanical-Responsive Explosion
Isolation Device

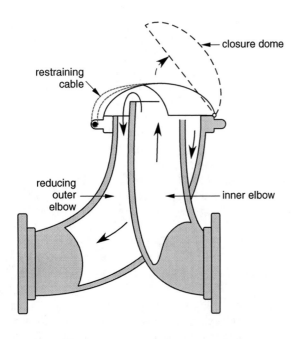

Figure 11 Dilution of Hydrocarbon Vapors

104. Dilution of Hydrocarbon Vapors

Dilution ventilation is the planned introduction of sufficient amounts of air to dilute unavoidable vapors to a nonhazardous level. The dilution ventilation approach is particularly useful with fugitive emissions, which are an intrinsic part of some manufacturing processes and, therefore, difficult to eliminate.

In order to properly calculate the dilution air requirements, the vapor production rate must be known. The goal is to keep the vapor density below one-quarter (i.e., 25%) of the lower explosion or flammability limit during periods of normal operations. The following steps can be used.

step 1: Determine the mass leak rates (e.g., in lbm/min or kg/min), $\dot{m}$, for each vapor component.

Note that many hydrocarbon-based liquids contain several pure hydrocarbons. Normally, data on each pure component are needed, and the dilution air for each is calculated separately. However, in order to simplify calculations, the total mass leak rate can be calculated by adding all component leak rates. In that case, the composite molecular weight must be used in subsequent calculations. This is determined by weighting the component molecular weights by the gravimetric fractions.

step 2: For each component, convert the mass leak rate into a volume leak rate (e.g., in ft³/min or ℓ/s) at the expected vapor temperature and pressure. This can be done in a variety of ways. One way is to use the ideal gas law with the mass leak rate, $\dot{m}$. The gas is assumed to be at atmospheric pressure.

$$\dot{V} = \frac{\dot{m}RT}{p} = \frac{\dot{m}R^*T}{p(\mathrm{MW})} \qquad 9$$

Alternatively, it can be recognized that 1 pound-mole (kg·mole) of any gas occupies 359 ft³ (10.2 m³) at standard scientific conditions. At the process temperature, the volume per mole would be

$$V_{\mathrm{pound\text{-}mole}} = (359\ \mathrm{ft^3})\left(\frac{T+460}{32+460}\right) \qquad \text{[U.S.] } 10(a)$$

$$V_{\mathrm{kg\cdot mole}} = (10.2\ \mathrm{m^3})\left(\frac{T+273}{273}\right) \qquad \text{[SI] } 10(b)$$

The volumetric generation rate would be found from

$$\dot{V} = \frac{\dot{m}V_{\mathrm{pound\text{-}mole}}}{(\mathrm{MW})} \qquad \text{[U.S.] } 11(a)$$

$$\dot{V} = \frac{\dot{m}V_{\mathrm{kg\text{-}mole}}}{(\mathrm{MW})} \qquad \text{[SI] } 11(b)$$

step 3: Determine the lower explosive/flammability limit for the vapor. Convert percentage LEL values to decimal fractions. Calculate the desired concentration as a decimal fraction by dividing by four.

$$C = \frac{\text{LEL}}{4} \qquad 12$$

step 4: Calculate the rate at which dilution gas (air) is introduced, Q, expressed in the same units as $\dot{V}$. Dilution air is the sum of outside fresh air and leaked gas.

$$C = \left(\frac{\dot{V}}{Q}\right)(1 - e^{-kn}) \qquad 13$$

k is the mixing efficiency factor, which ranges from 0.2 to 0.9, with the higher values implying the best conditions for mixing. n is the number of air changes that have occurred. However, once steady-state operation has been achieved, many air changes will have occurred. The last parenthetical term in Eq. 13 is essentially 1.0 after three to five air changes, and a value of 1.0 for that term can be used. Then,

$$Q = \frac{\dot{V}}{C_{\text{target}}} \qquad \text{[steady state]} \qquad 14$$

step 5: The intrinsic unreliability of the calculation process is recognized by multiplying the calculated air introduction rate by a safety factor of four.

$$Q' = 4Q \qquad 15$$

Example 13

Four pints of acetone (MW = 58.08, LEL = 2.6, SG = 0.792) evaporate each hour from a paint booth. The temperature of the acetone and paint booth is 80°F. What dilution air should be provided in order to avoid an explosive atmosphere?

(Solution)

step 1: The density of acetone is

$$\rho = (\text{SG})(\rho_{\text{water}}) = (0.792)(62.4 \text{ lbm/ft}^3)$$

$$= 49.42 \text{ lbm/ft}^3$$

The mass evaporation rate is

$$\dot{m} = \dot{V}\rho$$

$$= \frac{\left(4\,\frac{\text{pt}}{\text{hr}}\right)\left(0.1337\,\frac{\text{ft}^3}{\text{gal}}\right)\left(49.42\,\frac{\text{lbm}}{\text{ft}^3}\right)}{8\,\frac{\text{pt}}{\text{gal}}}$$

$$= 3.30 \text{ lbm/hr}$$

step 2: The approximate volume of a pound-mole of acetone vapor is given by Eq. 10.

$$\begin{aligned} V_{\text{pound-mole}} &= \frac{(359 \text{ ft}^3)(T + 460)}{32 + 460} \\ &= \frac{(359 \text{ ft}^3)(80 + 460)}{32 + 460} \\ &= 394 \text{ ft}^3 \end{aligned}$$

From Eq. 11, the volumetric rate at which vapors are generated is

$$\begin{aligned} \dot{V} &= \frac{\dot{m}V_{\text{pound-mole}}}{(\text{MW})} \\ &= \frac{\left(3.30\,\frac{\text{lbm}}{\text{hr}}\right)(394 \text{ ft}^3)}{58.08 \text{ lbm}} \\ &= 22.39 \text{ ft}^3/\text{hr} \end{aligned}$$

step 3: The lower explosive limit is 2.6% (0.026) by volume. The target concentration is

$$C = \frac{0.026}{4} = 0.0065$$

step 4: The dilution ventilation rate is given by Eq. 14.

$$Q = \frac{\dot{V}}{C} = \frac{22.39\,\frac{\text{ft}^3}{\text{hr}}}{0.0065}$$

$$= 3445 \text{ ft}^3/\text{hr}$$

step 5: Using a factor of safety of 4, the ventilation rate should be

$$Q' = 4Q = (4)\left(3445\,\frac{\text{ft}^3}{\text{hr}}\right)$$

$$= 13{,}780 \text{ ft}^3/\text{hr}$$

(Depending on the size of the paint booth, this ventilation rate may result in an excessively high air velocity.)

PART 6:
Practice Problems

1. A standpipe is being designed for a twin-bore tunnel subway, 50 feet underground. The slope of the tunnel is downward at 3%. The standpipe will be 500 feet long in each bore. Water will be supplied from a city water main at 100 psi on the surface. Assume a pipe friction factor of $f = 0.016$. Use NFPA 14.

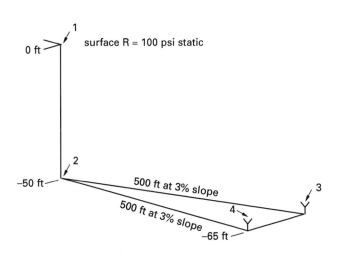

(a) What is the minimum flow required in the system?

(b) What are the minimum pipe sizes that can be used?

Answer

(a) Minimum flows are determined from NFPA 14, Sec. 2-1.2: 500 gpm in standpipe 3 and 250 gpm in standpipe 4. Thus, the total in the main is 750 gpm.

(b) The pipe sizing design criteria is based on having a 65 psi residual pressure at the hose valves 3

and 4 (NFPA 14, Sec. 2-1.1). Since the line is 500 feet long, the schedule pipe size from NFPA 14 is 6 inches. However, it is necessary to determine if the elevation drop can compensate for the pressure loss in the 500-foot line.

The cross-sectional areas of pipe are

$$A_{4''} = 0.088 \text{ ft}^2$$
$$A_{6''} = 0.20 \text{ ft}^2$$
$$A_{8''} = 0.35 \text{ ft}^2$$

The flow rates are

$$Q_{1\text{-}2} = \frac{750 \, \dfrac{\text{gal}}{\text{min}}}{\left(7.45 \, \dfrac{\text{gal}}{\text{ft}^3}\right)\left(60 \, \dfrac{\text{sec}}{\text{min}}\right)}$$
$$= 1.68 \text{ ft}^3/\text{sec}$$

$$Q_{2\text{-}3} = \frac{500 \, \dfrac{\text{gal}}{\text{min}}}{\left(7.45 \, \dfrac{\text{gal}}{\text{ft}^3}\right)\left(60 \, \dfrac{\text{sec}}{\text{min}}\right)}$$
$$= 1.12 \text{ ft}^3/\text{sec}$$

$$Q_{2\text{-}4} = \frac{250 \, \dfrac{\text{gal}}{\text{min}}}{\left(7.45 \, \dfrac{\text{gal}}{\text{ft}^3}\right)\left(60 \, \dfrac{\text{sec}}{\text{min}}\right)}$$
$$= 0.56 \text{ ft}^3/\text{sec}$$

Assume that the city water main is of sufficiently large capacity that the static and flowing water pressure at the source are the same.

Since a diameter of 6 inches is allowed by schedule, determine whether a 4-inch pipe will provide the minimum residual pressure at the specified flow.

From point 1 to point 2,

$$\frac{p_1}{\gamma} + \frac{v_1^2}{2g} + z_1 = \frac{p_2}{\gamma} + \frac{v_2^2}{2g} + z_2 + \frac{fLv_2^2}{2dg}$$

$$v_1 = 0 \quad \begin{bmatrix} \text{because static and flowing} \\ \text{pressure are the same} \end{bmatrix}$$

$$z_1 = 0 \quad [\text{datum}]$$

$$v_2 = \frac{Q_{1\text{-}2}}{A_{4''}} = \frac{1.68 \ \frac{\text{ft}^3}{\text{sec}}}{0.088 \ \text{ft}^2}$$

$$= 19.1 \ \text{ft/sec}$$

$$z_2 = -50 \ \text{ft}$$

$$f = 0.016$$

$$L = 50 \ \text{ft}$$

$$d = \frac{4 \ \text{in}}{12 \ \frac{\text{in}}{\text{ft}}} = 0.33 \ \text{ft}$$

$$p_1 = 100 \ \text{lbf/in}^2$$

$$\gamma = 62.4 \ \text{lbf/ft}^3$$

$$\left(\frac{100 \ \frac{\text{lbf}}{\text{in}^2}}{62.4 \ \frac{\text{lbf}}{\text{ft}^3}} \right) \left(144 \ \frac{\text{in}^2}{\text{ft}^2} \right) + 0 + 0$$

$$= \left(\frac{p_2 \ \frac{\text{lbf}}{\text{in}^2}}{62.4 \ \frac{\text{lbf}}{\text{ft}^3}} \right) \left(144 \ \frac{\text{in}^2}{\text{ft}^2} \right)$$

$$+ \left[\frac{\left(19.1 \ \frac{\text{ft}}{\text{sec}} \right)^2}{(2) \left(32.2 \ \frac{\text{ft}}{\text{sec}^2} \right)} \right] + (-50 \ \text{ft})$$

$$+ \left[\frac{(0.016)(50 \ \text{ft})}{0.33 \ \text{ft}} \right] \left[\frac{\left(19.1 \ \frac{\text{ft}}{\text{sec}} \right)^2}{(2) \left(32.2 \ \frac{\text{ft}}{\text{sec}^2} \right)} \right]$$

$$p_2(2.3 \ \text{ft}) = 230.8 \ \text{ft} - 5.7 \ \text{ft} + 50 \ \text{ft} - 13.7 \ \text{ft}$$

$$= 261.4 \ \text{ft}$$

$$p_2 = 113.7 \ \text{lbf/in}^2$$

From point 2 to point 3,

$$\frac{p_2}{\gamma} + \frac{v_2^2}{2g} + z_2 = \frac{p_3}{\gamma} + \frac{v_3^2}{2g} + z_3 + \frac{fLv_3^2}{2dg}$$

From the previous calculation,

$$\frac{p_2}{\gamma} = p_2(2.3 \ \text{ft}) = 261.4 \ \text{ft}$$

$$\frac{v_2^2}{2g} = 5.7 \ \text{ft}$$

$$z_2 = -50 \ \text{ft}$$

$$v_3 = \frac{Q_3}{A_4} = \frac{1.12 \ \frac{\text{ft}^3}{\text{sec}}}{0.088 \ \text{ft}^2}$$

$$= 12.7 \ \text{ft/sec}$$

$$z_3 = -50 \ \text{ft} + (500 \ \text{ft}) \left(-0.03 \ \frac{\text{ft}}{\text{ft}} \right)$$

$$= -65 \ \text{ft}$$

$$L = 500 \ \text{ft}$$

$$f = 0.016$$

$$d = 0.33 \ \text{ft}$$

$$261.4 \ \text{ft} + 5.7 \ \text{ft} + (-50 \ \text{ft})$$

$$= \left(\frac{p_3 \ \frac{\text{lbf}}{\text{in}^2}}{62.4 \ \frac{\text{lbf}}{\text{ft}^3}} \right) \left(144 \ \frac{\text{in}^2}{\text{ft}^2} \right)$$

$$+ \left[\frac{\left(12.7 \ \frac{\text{ft}}{\text{sec}} \right)^2}{(2)(32.2) \frac{\text{ft}}{\text{sec}^2}} \right] + (-65 \ \text{ft})$$

$$+ \left[\frac{(0.016)(500 \ \text{ft})}{0.33 \ \text{ft}} \right] \left[\frac{\left(12.7 \ \frac{\text{ft}}{\text{sec}} \right)^2}{(2)(32.2) \frac{\text{ft}}{\text{sec}^2}} \right]$$

$$261.4 \ \text{ft} + 5.7 \ \text{ft} - 50 \ \text{ft} = p_3(2.3 \ \text{ft}) + 2.5 \ \text{ft}$$
$$- 65 \ \text{ft} + 60.7 \ \text{ft}$$

$$p_3(2.3 \ \text{ft}) = 261.4 \ \text{ft} + 5.7 \ \text{ft} - 50 \ \text{ft}$$
$$- 2.5 \ \text{ft} + 65 \ \text{ft} - 60.7 \ \text{ft}$$

$$= 218.9 \ \text{ft}$$

$$p_3 = 95.2 \ \text{lbf/in}^2$$

This is greater than 65 psi, so a 4-inch line is ok.

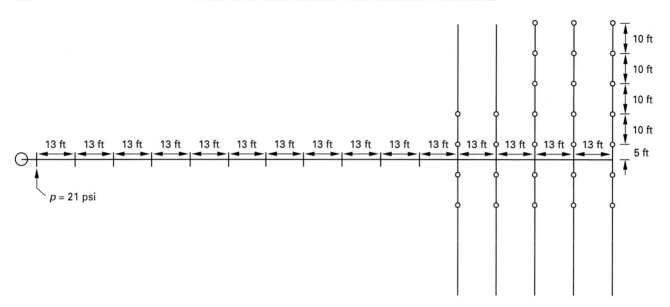

2. A sprinkler system is designed with 16 branches off a cross-main. The branches are 13 feet apart. Each branch is 45 feet long. The first sprinkler is 5 feet from the cross-main, and the remainder are spaced 10 feet apart. The hazard classification is ordinary 1. The design area is 1500 square feet. All sprinklers are $\frac{1}{2}$ inch with a K of 5.6. The friction loss between branches is limited to 0.5 psi.

(a) Determine the shape and outline of the design area.

(b) Assuming a minimum pressure of 21 psi at the start of the cross-main, what are the minimum sizes of pipe allowed in all 13-ft sections of the cross-main?

Answer

(a) The length of each branch is L_{branch}.

$$L_{\text{branch}} = 1.2\sqrt{1500 \text{ ft}^2}$$
$$= 46.5 \text{ ft}$$

The area each sprinkler covers is $A_{\text{sprinkler}}$.

$$A_{\text{sprinkler}} = (10 \text{ ft})(13 \text{ ft})$$
$$= 130 \text{ ft}^2$$

The spacing of sprinklers is $L_{\text{sprinkler}}$.

$$L_{\text{sprinkler}} = 10 \text{ ft}$$

The number of sprinklers on each branch is n_{branch}.

$$n_{\text{branch}} = \frac{L_{\text{branch}}}{L_{\text{sprinkler}}} = \frac{46.5 \text{ ft}}{10 \text{ ft}}$$
$$= 4.65$$
$$\approx 5 \text{ sprinklers/branch}$$

The total number of sprinklers is $n_{\text{sprinklers}}$.

$$n_{\text{sprinklers}} = \frac{\text{design area}}{A_{\text{sprinkler}}} = \frac{1500 \text{ ft}^2}{130 \text{ ft}^2}$$
$$= 11.5$$
$$\approx 12 \text{ sprinklers}$$

The total number of branches is n.

$$n = \frac{n_{\text{sprinklers}}}{n_{\text{branch}}} = \frac{12}{5}$$
$$= 2.4 \text{ branches}$$

$$(0.4 \text{ branches})\left(5 \; \frac{\text{sprinklers}}{\text{branch}}\right) = 2$$

Therefore, 2 sprinklers on branch 3. For tree systems these should be closest to the cross-main.

The following diagram shows the design area.

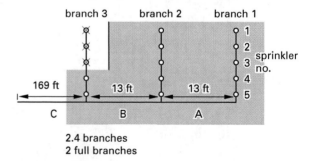

2.4 branches
2 full branches

(b) In order to determine the minimum cross-main sizes, the flows and pressure losses in the branches must be determined and then subtracted from the available pressure. The remainder is available for the cross-main. Since this is a wet system, $C = 120$. Therefore, the table in App. B can be used to calculate the pressure loss.

Sprinkler Worksheet

nozzle type and location	flow gpm (l/min)	pipe size (in)	fitting and devices	pipe equivalent length (ft)	friction loss psi/ft (bars/m)	required pressure psi (bars)
1	q 19.5 Q 19.5	1¼	length	10	0.032	p_t 12.1
			fittings	0		p_f 0.3
			total	10		p_e 0.0
2	q 19.7 Q 39.2	1¼	length	10	0.12	p_t 12.4
			fittings	0		p_f 1.2
			total	10		p_e 0.0
3	q 20.7 Q 59.9	2	length	10	0.035	p_t 13.6
			fittings	0		p_f 0.35
			total	10		p_e 0.0
4	q 21.0 Q 80.9	2	length	10	0.058	p_t 14.0
			fittings	0		p_f 0.58
			total	10		p_e 0.0
5	q 21.4 Q 102.3	2	length	5	0.095	p_t 14.6
			fittings	10		p_f 1.4
			total	15		p_e 0.0

The minimum flow at any sprinkler is 15 gpm (based on a 1/2-inch sprinkler at the minimum pressure of 7 psi.). However, check the flow required from the occupancy rating. From Fig. 9 for ordinary 1 occupancy, the minimum discharge density is 0.15 gpm/ft^2.

$$q_1 = \left(0.15 \, \frac{\text{gpm}}{\text{ft}^2}\right)(130 \text{ ft}^2) = 19.5 \text{ gpm}$$

$$q_1 = K\sqrt{p_{t,1}}$$

$$p_{t,1} = \left(\frac{q_1}{K}\right)^2 = \left(\frac{19.5 \text{ gpm}}{5.6}\right)^2 = 12.1 \text{ psi}$$

Disregard velocity pressure in all calculations. Thus, for no. 1, $q = 19.5$ gpm, and $p_t = 12.1$ psi.

From the chart in App. B, use $1\frac{1}{4}$-inch pipe. The loss is 0.032 psi/ft.

The pipe equivalent length is 10 feet.

Per NFPA 13, Sec. 6-4.4.5(d), the friction loss for the fitting directly connected to the sprinkler is excluded.

The total length is 10 feet.

The friction pressure loss, p_f, is (0.032 psi/ft)(10 ft) = 0.3 psi.

Because there is no change, the elevation pressure is $p_e = 0$.

Q is the branch total at this point, which is 19.5 gpm.

Sprinkler no. 2 is calculated similarly.

$$p_t = p_{t,1} + p_{f,1} = 12.1 \text{ psi} + 0.3 \text{ psi}$$

$$= 12.4 \text{ psi}$$

$$q = K\sqrt{p_t} = 5.6\sqrt{12.4} = 19.7 \text{ gpm}$$

$$Q = q + Q_1 = 19.7 \text{ gpm} + 19.5 \text{ gpm}$$

$$= 39.2 \text{ gpm}$$

$$\text{pipe size} = 1\tfrac{1}{4} \text{ in}$$

$$\text{length} = 10 \text{ ft}$$

$$\text{fittings} = 0 \text{ ft}$$

$$\text{total} = 10 \text{ ft} + 0 \text{ ft} = 10 \text{ ft}$$

$$\text{friction loss} = 0.12 \text{ psi/ft} \quad [\text{from App. B}]$$

$$p_f = (10 \text{ ft})(0.12 \text{ psi/ft}) = 1.2 \text{ psi}$$

$$p_e = 0.0 \text{ psi}$$

FIRE AND EXPLOSION PROTECTION SYSTEMS

Branch Section Worksheet

nozzle type and location	flow gpm (l/min)	pipe size (in)	fittings and devices		pipe equivalent length (ft)	friction loss psi/ft (bars/m)	required pressure psi (bars)	
A	q 102.3 Q 102.3	2½	length	13		0.038	p_t	16.0
			fittings	0			p_f	0.6
			total	13			p_e	0.0
B	q 102.3 Q 204.6	3½	length	13		0.025	p_t	16.5
			fittings	17			p_f	0.8
			total	30			p_e	0.0
C	q 46.6 Q 251.2	3½	length	169		0.025	p_t	15.0
			fittings	17			p_f	4.7
			total	186			p_e	0.0

Sprinkler no. 3:

$$p_t = p_{f,2} + p_{t,2} = 1.2 \text{ psi} + 12.4 \text{ psi}$$
$$= 13.6 \text{ psi}$$
$$q = K\sqrt{p_t} = 5.6\sqrt{13.6} = 20.7 \text{ gpm}$$
$$Q = q + Q_2 = 20.7 \text{ gpm} + 39.2 \text{ gpm}$$
$$= 59.9 \text{ gpm}$$

pipe size = 2 in

length = 10 ft

fittings = 0 ft

total = 10 ft + 0 ft = 10 ft

friction loss = 0.035 psi/ft

$$p_f = (10 \text{ ft})\left(0.035 \, \frac{\text{psi}}{\text{ft}}\right) = 0.35 \text{ psi}$$

$$p_e = 0.0 \text{ psi}$$

Sprinkler no. 4:

$$p_t = p_{t,3} + p_{f,3} = 13.6 \text{ psi} + 0.35 \text{ psi}$$
$$= 14.0 \text{ psi}$$
$$q = K\sqrt{p_t} = 5.6\sqrt{14.0} = 21.0 \text{ gpm}$$
$$Q = Q_3 + q = 59.9 \text{ gpm} + 21.0 \text{ gpm} = 80.9 \text{ gpm}$$

pipe size = 2 in

length = 10 ft

fittings = 0 ft

total = 10 ft + 0 ft = 10 ft

friction loss = 0.058 psi/ft

$$p_f = (10 \text{ ft})\left(0.058 \, \frac{\text{psi}}{\text{ft}}\right) = 0.58 \text{ psi}$$

$$p_e = 0.0 \text{ psi}$$

Sprinkler no. 5:

$$p_t = p_{t,4} + p_{f,4} = 14.0 \text{ psi} + 0.58 \text{ psi}$$
$$= 14.6 \text{ psi}$$
$$q = K\sqrt{p_t} = 5.6\sqrt{14.6}$$
$$= 21.4 \text{ gpm}$$
$$Q = Q_4 + q = 80.9 \text{ gpm} + 21.4 \text{ gpm}$$
$$= 102.3 \text{ gpm}$$

pipe size = 2 in

length = 5 ft

fittings = 10 ft   [tee, flow turned 90°]

total = 5 ft + 10 ft = 15 ft

friction loss = 0.25 psi/ft

$$p_f = (15 \text{ ft})\left(0.095 \, \frac{\text{psi}}{\text{ft}}\right) = 1.4 \text{ psi}$$

$$p_e = 0.0 \text{ psi}$$

Branch 2 is similar to branch 1. Branch 3 is similar to branch 1 for the first two sprinklers.

For section A, the pressure loss is limited to 0.5 psi. Thus the friction loss per foot is 0.5 psi/13 ft = 0.04 psi/ft. Using App. B, a 2½-inch pipe size is adequate. Disregard the friction from the upstream straight-through cross-fitting for branch 2. (The tee at the end of section A was included in nozzle 5.)

$$\text{total} = 13 \text{ ft} + 0 \text{ ft} = 13 \text{ ft}$$

The friction loss is 0.038 psi/ft.

$$p_t = p_{t,5} + p_{f,5} = 14.6 \text{ psi} + 1.4 \text{ psi}$$
$$= 16.0 \text{ psi}$$

$$p_f = \left(0.038 \ \frac{\text{psi}}{\text{ft}}\right)(13 \ \text{ft}) = 0.5 \ \text{psi} \qquad [\text{OK}]$$

$$p_e = 0.0 \quad [\text{no elevation change}]$$

For section B, the analysis is similar.

$$Q_B = q_B + Q_A = 102.3 \ \text{gpm} + 102.3 \ \text{gpm}$$
$$= 204.6 \ \text{gpm}$$

$$p_{t,B} = p_{t,A} + p_{f,A} = 16.0 \ \text{psi} + 0.5 \ \text{psi}$$
$$= 16.5 \ \text{psi}$$

From App. B, the required pipe size for section B is $3\frac{1}{2}$ inches. From App. D, the cross into branch 2 has an equivalent length of 17 feet. The friction loss per foot is 0.025 psi/ft.

$$p_f = \left(0.025 \ \frac{\text{psi}}{\text{ft}}\right)(13 \ \text{ft} + 17 \ \text{ft}) = 0.8 \ \text{psi}$$

$$p_e = 0.0 \ \text{psi}$$

In branch 3 the pressure at sprinklers 4 and 5 will be approximately the same as at the cross-main: The discharge from the two left-over sprinklers will be approximately

$$q_C = 2K\sqrt{p_t} = (2)(5.6)\sqrt{17.3} = 46.6 \ \text{gpm}$$
$$Q_C = q_C + Q_B = 46.6 \ \text{gpm} + 204.6 \ \text{gpm}$$
$$= 251.2 \ \text{gpm}$$

$$p_{t,C} = p_{t,B} + p_{f,B} = 16.5 \ \text{psi} + 0.8 \ \text{psi}$$
$$= 17.3 \ \text{psi}$$

The friction loss is estimated at

$$\frac{21 \ \text{psi} - 17.3 \ \text{psi}}{169 \ \text{ft} + 17 \ \text{ft}} = 0.02 \ \text{psi/ft}$$

From App. B, this requires a 4-in line.

$$p_f = \left(0.02 \ \frac{\text{psi}}{\text{ft}}\right)(186 \ \text{ft}) = 3.7 \ \text{psi}$$

$$p_{\text{required}} = 17.3 \ \text{psi} + 3.7 \ \text{psi}$$
$$= 21.0 \ \text{psi} \qquad [\leq 21 \ \text{psi available}]$$

Thus, a 4-inch cross-main is the minimum size usable down to the next-to-the-last branch, which is $3\frac{1}{2}$ inches, and to the last branch, which is $2\frac{1}{2}$ inches.

3. A sprinkler system is designed with 16 branches off two end cross-mains, with branches 13 feet apart. Each branch is 92 feet long. The first sprinklers are 1 foot from the cross-main, and the remainder are spaced 10 feet apart. The hazard classification is ordinary 1. The design area is 1500 square feet. All sprinklers are $\frac{1}{2}$ inch with a K of 5.6. The friction loss between branches is limited to 0.5 psi.

(a) Determine the shape and outline of the design area.

(b) Assuming a minimum pressure of 20 psi at the start of the cross-mains, what are the minimum pipe sizes allowed in the cross-mains?

Answer

(a) The length of each branch in the design area is L_{branch}.

$$L_{\text{branch}} = 1.2\sqrt{1500 \ \text{ft}^2} = 46.5 \ \text{ft}$$

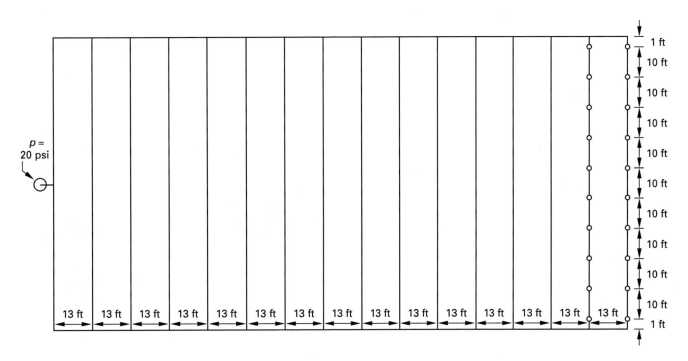

Sprinkler Worksheet

nozzle type and location	flow gpm (l/min)	pipe size (in)	fittings and devices		pipe equivalent length (ft)	friction loss psi/ft (bars/m)	required pressure psi (bars)	
3, 4	q 19.5 Q 19.5	1¼	length fittings total	10 0 10		0.032	p_t 12.1 p_f 0.3 p_e 0.0	
2, 5	q 19.7 Q 39.2	1¼	length fittings total	10 0 10		0.12	p_t 12.4 p_f 1.2 p_e 0.0	
1, 6	q 20.7 Q 59.9	2	length fittings total	21 5 26		0.035	p_t 13.6 p_f 0.9 p_e 14.5	

Each sprinkler covers an area, $A_{sprinkler}$.

$$A_{sprinkler} = (10 \text{ ft})(13 \text{ ft}) = 130 \text{ ft}^2$$

The spacing between sprinklers, $L_{sprinkler}$, is 10 ft.

The number of sprinklers on each branch is n_{branch}.

$$n_{branch} = \frac{L_{branch}}{L_{sprinkler}}$$

$$= \frac{46.5 \text{ ft}}{10 \text{ ft}} = 4.65$$

$$\approx 5 \text{ sprinklers/branch}$$

Note that this is a minimum and may be increased to six sprinklers for symmetrical analysis.

The total number of sprinklers is $n_{sprinklers}$.

$$n_{sprinklers} = \frac{\text{design area}}{A_{sprinkler}}$$

$$= \frac{1500 \text{ ft}^2}{130 \text{ ft}^2} = 11.5$$

$$\approx 12 \text{ sprinklers}$$

The total number of branches is n.

$$n = \frac{n_{sprinklers}}{n_{branch}} = \frac{12}{5}$$

$$= 2.5 \text{ branches}$$

Use $n_{branch} = 6$.

$$n = \frac{12}{6} = 2 \text{ branches}$$

The following diagram shows the design area.

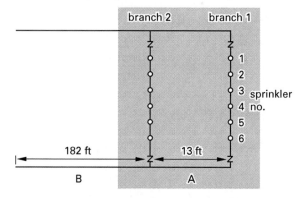

(b) To determine the minimum cross-main sizes, the flows and pressure losses in the branches must be determined and then subtracted from the available pressure. The remainder is available for the cross-main. Since this is a wet system, $C = 120$. The table in App. B can be used to calculate pressure loss. Since the sprinkler arrangement is symmetrical, it can be assumed that sprinklers 1, 2, and 3 are fed from one side, and sprinklers 4, 5, and 6 are fed from the other.

The minimum flow at any sprinkler must be 15 gpm. However, check the flow required from the occupancy rating. From Fig. 9 for ordinary 1 occupancy, the minimum discharge density is 0.15 gpm/ft².

$$q_1 = \left(0.15 \; \frac{\text{gpm}}{\text{ft}^2}\right) (130 \text{ ft}^2) = 19.5 \text{ gpm}$$

$$q_{3,4} = K\sqrt{p_{t,3,4}}$$

$$p_{t,3,4} = \left(\frac{q_{3,4}}{K}\right)^2 = \left(\frac{19.5}{5.6}\right)^2$$

$$= 12.1 \text{ psi}$$

Branch Section Worksheet

nozzle type and location	flow gpm (l/min)	pipe size (in)	fittings and devices		pipe equivalent length (ft)	friction loss psi/ft (bars/m)	required pressure psi (bars)	
A	q 59.9	2	length	13		0.035	p_t 14.5	
			fittings	0			p_f 0.5	
	Q 59.9		total	13			p_e 0.0	
B	q 50	2½	length	182		0.038	p_t 13.5	
			fittings	12			p_f 7.8	
	Q 100		total	194			p_e 0.0	

Disregard velocity pressure in all calculations.

Thus, $q_{3,4} = 19.5$ gpm and $p_{t,3,4} = 12.1$ psi.

From the chart in App. B, use $1\frac{1}{4}$-inch pipe. The loss in psig/ft $= 0.032$. The pipe equivalent length is 10 feet (NFPA 13, Sec. 6-4.4.5(a)). The friction loss for the fitting directly connected to the sprinkler is excluded.

The total length is 10 feet.

The friction pressure is $p_{f,3,4}$.

$$p_{f,3,4} = \left(0.032 \, \frac{\text{psi}}{\text{ft}}\right)(10 \text{ ft}) = 0.3 \text{ psi}$$

Since there is no change in elevation, $p_{e,3,4} = 0.0$ psi.

$Q_{3,4}$ is the branch total, which is the same as $q_{3,4}$.

For sprinklers 2 and 5,

$$p_{t,2,5} = p_{t,3,4} + p_{f,3,4} = 12.1 \text{ psi} + 0.3 \text{ psi}$$
$$= 12.4 \text{ psi}$$
$$q_{2,5} = K\sqrt{p_{t,2,5}} = 5.6\sqrt{12.4}$$
$$= 19.7 \text{ gpm}$$
$$Q_{2,5} = q_{2,5} + Q_{3,4}$$
$$= 19.5 \text{ gpm} + 19.7 \text{ gpm}$$
$$= 39.2 \text{ gpm}$$
$$\text{pipe size} = 1\frac{1}{4} \text{ in}$$
$$\text{friction loss} = 0.12 \text{ psi/ft}$$

$$p_{f,2,5} = \left(0.12 \, \frac{\text{psi}}{\text{ft}}\right)(10 \text{ ft})$$
$$= 1.2 \text{ psi}$$
$$p_{e,2,5} = 0.0 \text{ psi} \quad [\text{no elevation change}]$$

For sprinklers 1 and 6,

$$p_{t,1,6} = p_{t,2,5} + p_{f,2,5} = 12.4 \text{ psi} + 1.2 \text{ psi}$$
$$= 13.6 \text{ psi}$$
$$q_{1,6} = K\sqrt{p_{t,1,6}} = 5.6\sqrt{13.6}$$
$$= 20.7 \text{ gpm}$$
$$Q_{1,6} = Q_{2,5} + q_{1,6} = 39.2 \text{ gpm} + 20.7 \text{ gpm}$$
$$= 59.9 \text{ gpm}$$
$$\text{length} = 2 \times 10 \text{ ft} + 1 \text{ ft}$$
$$= 21 \text{ ft} \quad \begin{bmatrix} \text{sprinkler 6} \\ \text{to end branch} \end{bmatrix}$$
$$\text{pipe size} = 2 \text{ in}$$
$$\text{fittings} = 5 \text{ ft} \quad [\text{standard elbow}]$$
$$\text{total} = 21 \text{ ft} + 5 \text{ ft} = 26 \text{ ft}$$
$$\text{friction loss} = 0.035 \text{ psi/ft}$$
$$p_{f,6} = \left(0.035 \, \frac{\text{psi}}{\text{ft}}\right)(26 \text{ ft}) = 0.9 \text{ psi}$$
$$p_{e,1,6} = 0.0 \text{ psi} \quad [\text{because there is no change}]$$

Branch 2 is similar to branch 1, except that a cross is used instead of an elbow.

For cross-main section A, the pressure loss is limited to 0.5 psi. Thus, the friction loss per foot is 0.5 psi/13 ft, which is 0.04 psi/ft. Using App. B, a 2-inch pipe is adequate.

$$q_A = Q_{1,6} = 59.9 \text{ gpm}$$

$$Q_A = q_A$$

$$\text{length} = 13 \text{ ft}$$

$$\text{fittings} = 0 \quad \text{[included in nozzle 1-6]}$$

$$\text{total length} = 13 \text{ ft} + 0 = 13 \text{ ft}$$

$$p_{t,A} = p_{t,1,6} + p_{f,1,6} = 13.6 \text{ psi} + 0.9 \text{ psi}$$

$$= 14.5 \text{ psi}$$

$$\text{friction loss} = 0.035 \text{ psi/ft}$$

$$p_{f,A} = \left(0.035 \, \frac{\text{psi}}{\text{ft}}\right)(13 \text{ ft}) = 0.5 \text{ psi}$$

$$p_{e,A} = 0.0 \quad \text{[because there is no change]}$$

For cross-main B the flow rate is essentially doubled.

$$Q_B = q_A + Q_A = 59.9 \text{ gpm} + 59.9 \text{ gpm}$$

$$= 119.8 \text{ gpm}$$

$$\text{length} = 182 \text{ ft}$$

$$\text{fittings} = 12 \text{ ft} \quad \begin{bmatrix} \text{cross into branch 2;} \\ \text{assuming } 2\frac{1}{2}\text{-in pipe} \end{bmatrix}$$

$$p_{t,B} = p_{t,A} + p_{f,A} = 14.5 \text{ psi} + 0.5 \text{ psi}$$

$$= 15 \text{ psi}$$

$$\begin{matrix} \text{estimated} \\ \text{friction} \\ \text{loss} \end{matrix} = \frac{20 \text{ psi} - 15 \text{ psi}}{182 \text{ ft} + 12 \text{ ft}}$$

$$= 0.026 \text{ psi/ft}$$

This implies a 3-inch pipe. Actual loss is approximately 0.02 psi/ft. Actual cross-fitting loss is 15 ft.

$$p_{f,B} = \left(0.026 \, \frac{\text{psi}}{\text{ft}}\right)(182 \text{ ft} + 15 \text{ ft}) = 5.1 \text{ psi}$$

$$p_{e,B} = 0.0$$

$$p_{\text{required}} = p_{t,B} + p_{f,B} = 15 \text{ psi} + 5.1 \text{ psi}$$

$$= 20.1 \text{ psi} \quad [> 20 \text{ psi available}]$$

4. A standpipe and sprinkler system are being designed for a building. The system requires 1000 gpm at 20 psi at the most hydraulically remote outlet. A test is conducted on a municipal utility hydrant located next to the building. The following data is recorded:

- static pressure of 30 psi

- flow at 800 gpm, residual pressure of 22 psi

(a) Is a storage tank required for a 60-minute fire water supply?

(b) If a tank is required, what should be its capacity to the nearest 1000 gal?

Answer

(a) A storage tank is required if the design flow cannot be maintained with a 20 psi residual pressure on the utility line. Using the Hazen-Williams equation as a basis, the friction loss per foot is defined as $p_{f/\text{ft}}$.

$$p_{f/\text{ft}} = \frac{4.52 Q_{\text{gpm}}^{1.85}}{C^{1.85} d_{\text{in}}^{4.87}}$$

For a given system, all of the values except Q are constants. In addition, the friction loss is a pressure differential, with $p_f = 0$ when $Q = 0$. Thus, the equation can be written as follows:

$$\Delta p_f = K Q_{\text{gpm}}^{1.85}$$

For any two flows, 1 and 2,

$$\frac{\Delta p_{f,1}}{Q_{\text{gpm},1}^{1.85}} = \frac{\Delta p_{f,2}}{Q_{\text{gpm},2}^{1.85}} = K$$

For a given hydrant flow test, the system K is fixed. The pressure drop for other flows can be calculated.

Define h as Δp_f; the subscript f represents the measured test results, and subscript r is the value to be calculated. (This calculation method is also given in NFPA Recommended Practice 291, Fire Flow Testing and Marking of Hydrants.)

$$\frac{h_r}{Q_r^{1.85}} = \frac{h_f}{Q_f^{1.85}}$$

$$Q_r = Q_f \left(\frac{h_r^{0.54}}{h_f^{0.54}}\right)$$

$$h_f = 30 \text{ psi} - 22 \text{ psi} = 8 \text{ psi}$$

$$Q_f = 800 \text{ gpm}$$

$$Q_r = 1000 \text{ gpm}$$

Find h_r.

$$h_r = h_f \left(\frac{Q_r}{Q_f}\right)^{1.85}$$

$$= 8 \left(\frac{1000 \text{ gpm}}{800}\right)^{1.85} = 12.1 \text{ psi}$$

The utility pressure at 1000 gpm would be 30 psi − 12.1 psi = 17.9 psi, which is less than 20 psi; therefore, a storage tank is required.

(b) The maximum flow that can be delivered from the municipal supply is Q_r.

$$h_f = 30 \text{ psi} - 22 \text{ psi} = 8 \text{ psi}$$
$$Q_f = 800 \text{ gpm}$$
$$h_r = 30 \text{ psi} - 20 \text{ psi} = 10 \text{ psi}$$
$$Q_r = Q_f \left(\frac{h_r}{h_f}\right)^{0.54} = (800 \text{ gpm})\left(\frac{10 \text{ psi}}{8 \text{ psi}}\right)^{0.54}$$
$$= 902 \text{ gpm}$$

Thus, the storage tank should make up the flow rate difference for the required duration.

$$Q_{\text{tank}} = (1000 \text{ gpm} - 902 \text{ gpm})(60 \text{ min})$$
$$= 5880 \text{ gal}$$
$$\approx 6000 \text{ gal} \quad \text{[to the nearest 1000 gal]}$$

5. A dry standpipe is to be filled with two fire pumps in parallel, each rated at 500 gpm and 65 psi. The standpipe is located in one side of a 300-foot tunnel. The line slopes 3% downward. The pump inlet is located 40 feet below the surface of a large reservoir. All piping is 6-inch carbon steel. Assume 10% over measured length for fitting loss and a Darcy friction factor of 0.032. The fire department has stipulated that the calculated fill rate without the pumps cannot exceed the maximum flow of both fire pumps.

(a) Is the fire department stipulation on flow rate a limitation on the system fill time?

(b) What is the estimated fill time of the dry pipeline without the pumps operating?

Answer

(a) Determine the fire pumps' performance curve. Since the pumps are in parallel, their flows are combined for each point. Using NFPA 20, Secs. 3-21 and 4-13, three limiting conditions apply which which can be extrapolated to estimate a maximum pump output.

flow (gpm)	percent of design (%)	pressure (psi)
0	140	91
500	100	65
750	67	44

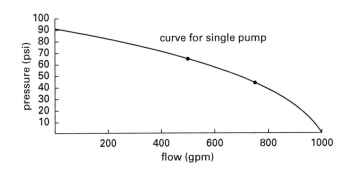

Thus, the maximum flow from the two pumps is 2000 gpm at 0 psi.

Determine the system flow without the pump.

Define the reservoir surface as point 1 and the end of the pipeline as point 2. Find the flow rate at the

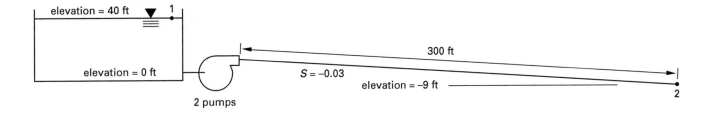

end of the pipeline, assuming steady flow. Writing the energy equation,

$$\frac{p_1}{\gamma} + \frac{v_1^2}{2g} + z_1 = \frac{p_2}{\gamma} + \frac{v_2^2}{2g} + z_2 + \frac{fL_e v_2^2}{2dg}$$

$p_1 = p_2 = 0$ [atmospheric pressure]

$v_1 = 0$ $\left[\begin{array}{c}\text{velocity of reservoir much} \\ \text{lower than system}\end{array}\right]$

$z_1 = 40$ ft [elevation of reservoir]

$z_2 = 0$ ft $-$ (300 ft)(-0.03)

$\quad = -9$ ft

$f = 0.032$

$L_e = (300\text{ ft})(1.1) = 330$ ft

$d = \dfrac{6.065\text{ in}}{12\ \frac{\text{in}}{\text{ft}}} = 0.51$ ft

Substituting,

$$0 + 0 + 40\text{ ft} = 0 + \frac{v_2^2}{2g} + (-9\text{ ft})$$

$$+ \left[\frac{(0.032)(330\text{ ft})}{0.51\text{ ft}}\right]\left(\frac{v_2^2}{2g}\right)$$

$$40\text{ ft} - (-9\text{ ft}) = \left(\frac{v_2^2}{2g}\right)\left[1 + \frac{(0.032)(330\text{ ft})}{0.51\text{ ft}}\right]$$

$$49\text{ ft} = \left(\frac{v_2^2}{2g}\right)(1 + 20.7)$$

$$= \sqrt{\frac{(2)\left(32.2\ \frac{\text{ft}}{\text{sec}^2}\right)(49\text{ ft})}{21.7}}$$

$$= 12.1\text{ ft/sec}$$

The flow rate, Q, is $v_2 A$, where A is the area of the 6-inch pipe.

$$Q = v_2 A = v_2 \left(\frac{\pi}{4}\right)d^2$$

$$= \left(12.1\ \frac{\text{ft}}{\text{sec}}\right)\left(\frac{\pi}{4}\right)(0.51\text{ ft})^2$$

$$\times \left(7.45\ \frac{\text{gal}}{\text{ft}^3}\right)\left(60\ \frac{\text{sec}}{\text{min}}\right)$$

$$= 1105\text{ gpm}$$

Since 1105 gpm is the smallest flow rate and this does not exceed the combined pump capacity of 2000 gpm, the pumps are not the limiting factor. Thus, the fire department stipulation is not a limitation on the flow rate to the system.

(b) The estimated fill time, t, is L/v, where L is the length of the line and v is the velocity.

$$L = 300\text{ ft}$$

$$v = \frac{Q}{A}$$

$$= \frac{1105\ \frac{\text{gal}}{\text{min}}}{\left(7.45\ \frac{\text{gal}}{\text{ft}^3}\right)\left(60\ \frac{\text{sec}}{\text{min}}\right)\left(\frac{\pi}{4}\right)(0.51\text{ ft})^2}$$

$$= 12.10\text{ ft/sec}$$

$$t = \frac{L}{v} = \frac{300\text{ ft}}{12.10\ \frac{\text{ft}}{\text{sec}}}$$

$$= 24.8\text{ sec}$$

6. A building is 200 feet long by 100 feet wide. Structural considerations dictate branch line spacing of 13 feet. Cross-mains are located at both 200-foot walls and run to a common riser on one end. The occupancy hazard classification of the building is ordinary 1. The design area has been designated as 1500 square feet.

(a) Determine the design area using the area/density method.

(b) Determine the required demand for sprinklers.

Answer

(a) Per NFPA 13, Sec. 6-4.4.1, the design area using the area/density method is based on a rectangle having a dimension parallel to the branch line of at least 1.2 multiplied by the square root of the design area. The design area must be the most hydraulically demanding area of the system.

The length of each branch is L_{branch}.

$$L_{\text{branch}} = 1.2\sqrt{1500\text{ ft}^2} = 46.5\text{ ft}$$

Each sprinkler covers an area $A_{\text{sprinkler}}$.

$$A_{\text{sprinkler}} = (10\text{ ft})(13\text{ ft}) = 130\text{ ft}^2$$

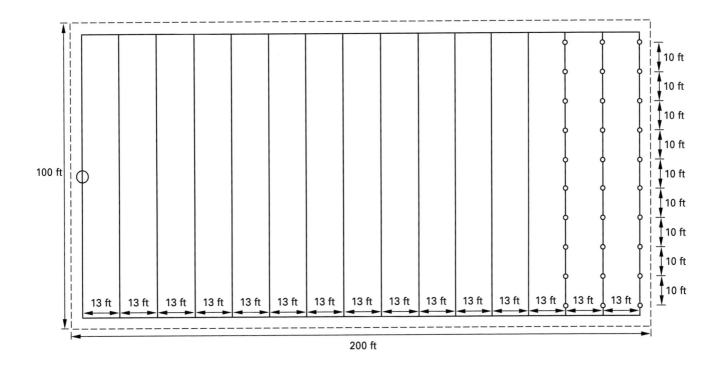

The number of sprinklers on each branch is n_{branch}.

$$L_{\text{sprinkler}} = 10 \text{ ft}$$

$$n_{\text{branch}} = \frac{L_{\text{branch}}}{L_{\text{sprinkler}}} = \frac{46.5 \text{ ft}}{10 \text{ ft}}$$

$$= 4.65$$

$$\approx 5 \text{ sprinklers/branch}$$

The total number of sprinklers is $n_{\text{sprinklers}}$.

$$n_{\text{sprinklers}} = \frac{\text{design area}}{A_{\text{sprinkler}}} = \frac{1500 \text{ ft}^2}{\dfrac{130 \text{ ft}^2}{\text{sprinkler}}}$$

$$= 11.5 \text{ sprinklers}$$

$$\approx 12 \text{ sprinklers}$$

The total number of branches is n.

$$n = \frac{n_{\text{sprinkler}}}{n_{\text{branch}}} = \frac{12}{5}$$

$$= 2.4 \text{ branches}$$

Since this must be the hydraulically remote area, it must be centered between the two cross-mains.

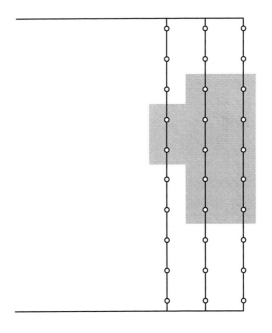

(b) The minimum flow required is Q_{total}. The design density for 1500 ft² of area with ordinary 1 classification is 0.15 gpm/ft². There are 12 sprinklers, each covering 130 ft².

$$Q_{\text{total}} = (12)(130 \text{ ft}^2)\left(0.15 \ \frac{\text{gpm}}{\text{ft}^2}\right)$$

$$= 234 \text{ gpm}$$

7. The building in problem 6 is reconfigured as a food service court with a central 50-foot-wide walkway and eating area extending the full length. The kitchen and preparation area are on both sides of the walkaway and are 25 feet wide and 40 feet long. The eating area is classified as light occupancy hazard, and the design area is 3000 square feet. The kitchen and preparation areas are classified as ordinary 1.

(a) Determine the design areas.

(b) Based on sprinkler demand, are changes to the sprinkler system necessary?

Answer

(a) For design area 1 (the seating in center of the building) from the area density curve for light hazard, the density is 0.075 gpm/ft² for a 3000-ft² design area. Thus, demand is given by Q.

$$Q_1 = \left(0.075 \ \frac{\text{gpm}}{\text{ft}^2}\right)(3000 \ \text{ft}^2) = 225 \ \text{gpm}$$

Because the design area is less than the total area, use the area/density method.

$$L_{\text{branch}} = 1.2\sqrt{3000 \ \text{ft}^2} = 65.7 \ \text{ft}$$
$$L_{\text{sprinkler}} = 10 \ \text{ft}$$

$$n_{\text{branch}} = \frac{L_{\text{branch}}}{L_{\text{sprinkler}}} = \frac{65.7 \ \text{ft}}{10 \ \text{ft}}$$
$$= 6.6 \ \text{sprinklers}$$

Since only six sprinklers are available, use them all and extend the number of branches. (NFPA 13, Sec. 6-4.4.1a, Exception 1.)

$$n_{\text{sprinklers}} = \frac{3000 \ \text{ft}^2}{130 \ \text{ft}^2} = 23.1$$
$$\approx 24$$
$$n = \frac{24}{6}$$
$$= 4 \ \text{branches}$$

This is shown as follows.

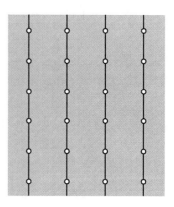

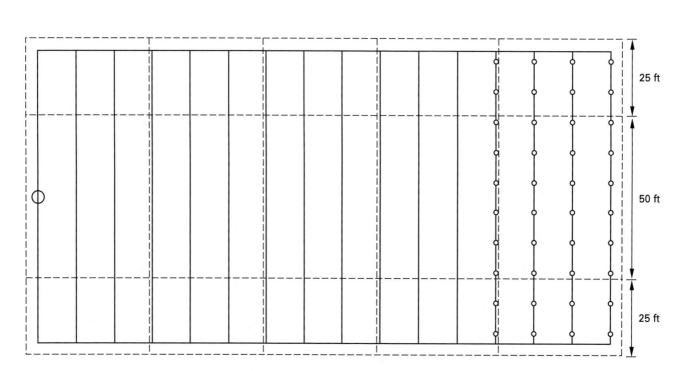

Next, work with design area 2, the kitchen and preparation area.

Since this is an isolated room, use the room design method. From the area density curve for ordinary hazard, the density is 0.15 gpm/ft². Demand is given by Q_2.

$$Q_2 = \left(0.15 \ \frac{\text{gpm}}{\text{ft}^2}\right)(25 \ \text{ft})(40 \ \text{ft})$$
$$= 150 \ \text{gpm}$$

The design area is the entire room.

(b) The original design flow was 225 gpm.

$$Q_1 = 225 \ \text{gpm} \quad \text{[same]}$$
$$Q_2 = 150 \ \text{gpm} \quad \text{[less]}$$

Thus, no change is necessary to the sprinkler system based on hydraulic demand. (Some relocation of individual sprinklers may be necessary to meet spacing requirements.)

8. A person has fallen in a 2000 ft³ storage tank and will bleed to death in 5 minutes unless rescue workers can stop the bleeding. An explosion-proof ventilating fan will not be available for 10 minutes. When the worker falls in, a gas line begins delivering 1700 ft³/min of methane gas. It is estimated that atmospheric leakage from open hatches and so on is 500 ft³/min. It is estimated that SCBA-equipped rescue workers will enter in 1 minute.

(a) Can a non-explosive atmosphere be maintained for the period of time needed for rescue workers to reach the victim and stop the bleeding?

(b) What is the factor of safety when the rescue workers first enter?

Answer

The conditions are as follows. Assume the vented gases are poorly mixed.

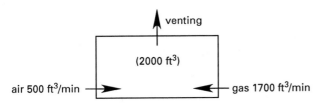

(a) The lower explosive limit for methane is 5%, and the upper explosive limit is 15% (App. E). Determine whether the rate of gas input is sufficient

to maintain the concentration above 15% with the venting taking place, where

$$C = \left(\frac{\dot{V}_{\text{gas}}}{Q_{\text{air}}}\right)\left(1 - e^{-kn}\right)$$

Determine the number of air changes, n, after 1 minute. Use $k = 0.2$ because the gases are poorly mixed.

$$n = \frac{\left(1700 \ \dfrac{\text{ft}^3}{\text{min}} + 500 \ \dfrac{\text{ft}^3}{\text{min}}\right)(1 \ \text{min})}{2000 \ \text{ft}^3}$$
$$= 1.1$$

$$C = \left(\frac{1700 \ \dfrac{\text{ft}^3}{\text{min}}}{500 \ \dfrac{\text{ft}^3}{\text{min}} + 1700 \ \dfrac{\text{ft}^3}{\text{min}}}\right)\left(1 - e^{-(0.2)(1.1)}\right)$$
$$= 0.153 \quad [>0.15; \text{ safe}]$$

This concentration is outside the 5%–15% explosive range.

(b) The safety factor is

$$S = \frac{C}{\text{UEL}} = \frac{0.153}{0.15} = 1.0$$

The factor of safety will increase with time.

9. A robotic automotive paint bay is being operated continuously. Up to 1 gallon of toluene (toluol) thinner (molecular weight = 92.13, and specific gravity = 0.866) may be sprayed in the bay per hour. The operating temperature is 80°F. What is the design ventilation requirement in ft³/min?

Answer

Since this is an automated paint booth, it can be assumed that toxicity dilution is not a concern. Thus, evaluation need only be made for explosive level dilution. From App. E, for toluene, LEL = 1.3% by volume.

The density of toluene is

$$\rho = (\text{SG})(\rho_{\text{water}}) = (0.866)\left(62.4 \ \frac{\text{lbm}}{\text{ft}^3}\right)$$
$$= 54.04 \ \text{lbm/ft}^3$$

The mass evaporation rate is

$$\dot{m} = \dot{V}\rho$$
$$= \left(1 \ \frac{\text{gal}}{\text{hr}}\right)\left(0.1337 \ \frac{\text{ft}^3}{\text{gal}}\right)\left(54.04 \ \frac{\text{lbm}}{\text{ft}^3}\right)$$
$$= 7.23 \ \text{lbm/hr}$$

The approximate volume of a lbm-mole of vapor is

$$V_{\text{lbm-mole}} = \frac{(359 \text{ ft}^3)(T + 460°F)}{32°F + 460°F}$$

$$= \frac{(359 \text{ ft}^3)(80°F + 460°F)}{32°F + 460°F}$$

$$= 394 \text{ ft}^3$$

The volumetric rate at which vapors are generated is

$$\dot{V} = \frac{\dot{m} V_{\text{lbm-mole}}}{\text{MW}}$$

$$= \frac{\left(7.23 \dfrac{\text{lbm}}{\text{hr}}\right)\left(394 \dfrac{\text{ft}^3}{\text{lbm-mole}}\right)}{92.13 \dfrac{\text{lbm}}{\text{lbm-mole}}}$$

$$= 30.92 \frac{\text{ft}^3}{\text{hr}}$$

The lower explosive limit is 1.3%. The target concentration is 25% of this.

$$C = (0.013)(0.25) = 0.0032$$

The dilution ventilation rate in ft^3/min is

$$Q = \frac{\dot{V}}{C} = \frac{\left(30.92 \dfrac{\text{ft}^3}{\text{hr}}\right)}{(0.0032)\left(60 \dfrac{\text{min}}{\text{hr}}\right)}$$

$$= 161 \text{ ft}^3/\text{min}$$

10. A turbine enclosure measuring 20 feet long, 10 feet wide, and 10 feet high with an effective air inlet of 5 ft^2 is to be protected by a total-flooding carbon dioxide system. Charts show that the volume factor for 34% carbon dioxide concentration is 18 ft^3 per pound of carbon dioxide. Charts also show that the leakage rate for the enclosure is 17 lbm of carbon dioxide/ft^2 of opening per minute. It is desired to keep a 34% carbon dioxide concentration inside the enclosure for 1 minute. What is the mass (in lbm) of carbon dioxide needed?

Answer

The enclosure volume is

$$V = (10 \text{ ft})(10 \text{ ft})(20 \text{ ft}) = 2000 \text{ ft}^3$$

The mass to fill the volume is m_{base}.

$$m_{\text{base}} = \frac{2000 \text{ ft}^3}{18 \dfrac{\text{ft}^3}{\text{lbm}}} = 111 \text{ lbm}$$

The mass lost through the openings is m_{leak}.

$$m_{\text{leak}} = (5 \text{ ft}^2)\left(17 \frac{\text{lbm}}{\text{ft}^2\text{-min}}\right)(1 \text{ min})$$

$$= 85 \text{ lbm}$$

The total mass of CO_2 required is the sum of $m_{\text{base}} + m_{\text{leak}}$.

$$m_{\text{base}} + m_{\text{leak}} = 111 \text{ lbm} + 85 \text{ lbm}$$

$$= 196 \text{ lbm}$$

APPENDICES

APPENDIX A

Typical Thermal Properties of
Selected Building Materials[a]
(English units)

material	density $\left(\dfrac{\text{lbm}}{\text{ft}^3}\right)$	thermal conductivity $\left(\dfrac{\text{BTU-in}}{\text{hr-ft}^2\text{-}°\text{F}}\right)$	specific heat $\left(\dfrac{\text{BTU}}{\text{lbm-}°\text{F}}\right)$	increase in length for each 100°F temp rise (%)	melting point (°F)
air	0.06	0.2	0.24		
water	62	5	1.0	0.01	32
aluminum	165	1400	0.22	0.14	1220
brass	530	720	0.09	0.11	1650
copper	560	2600	0.09	0.09	1980
cast iron	440	320	0.13	0.06	2466–2550
steel	490	310	0.12	0.06–0.15	2370–2550
glass	160	6	0.20	0.04–0.06	2600
brick	120	5	0.22	0.05	–
concrete, normal weight	140	9–12	0.16–0.25	0.06–0.08	–
asbestos-cement board	120	4	0.2		
wood (oak, maple)	45	1.1	0.30–0.55	0.03–0.05	–
wood (fir, pine)	32	0.8	0.33–0.45	0.02–0.03	–
hardboard	65	1.0	0.33	–	–
plywood	35	0.8	0.29	–	–
fiberboard (wood or cane)	15	0.35	0.30	–	–
plaster	70	3–6	0.23	–	–
gypsum board	50–60	1.5	0.26	–	–
glass fiber batt	0.6	0.5	0.2	–	–
	3	0.3	0.2	–	–
mineral wool	3	0.3	0.25	–	–
plastics, rigid					
vinyls	–	1–2	0.2–0.3	0.3–1.0	–
styerene	–	0.7–1.0	0.32–0.35	0.3–0.4	–
polystyrene foam	2	0.26	0.32	0.3–0.4	–
polyurethane foam	2	0.18	0.38	0.4	–

Multiply lbm/ft^3 by 16.018 to obtain kg/m^3.
Multiply BTU-in/ft^2-hr-°F by 0.14423 to obtain W/m·K.
Multiply BTU/lbm-°F by 4186.8 to obtain J/kg-K.
Use $\left(\frac{5}{9}\right)(°\text{F}-32)$ to calculate temperature in °C.

[a]Values listed are estimated values at ordinary temperatures or over typical temperature ranges in fires, if available. Actual values vary considerably with temperature, particularly where moisture is involved.

Reprinted from *Fire Investigation Handbook*, National Bureau of Standards Handbook 134, 1980.

APPENDIX B

Friction Loss in Schedule-40 Steel Pipe (English units)

(Hazen-Williams $C = 120$)[a]

loss in psig/ft

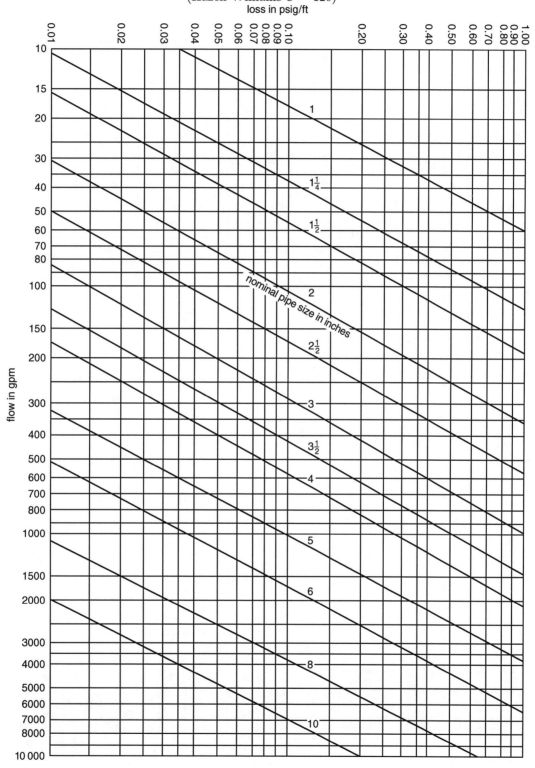

[a] Use Table 14 to convert to other C-values.

APPENDIX C

Friction Loss in Schedule-40 Steel Pipe (SI Units)
(Hazen-Williams $C = 120$)[a]

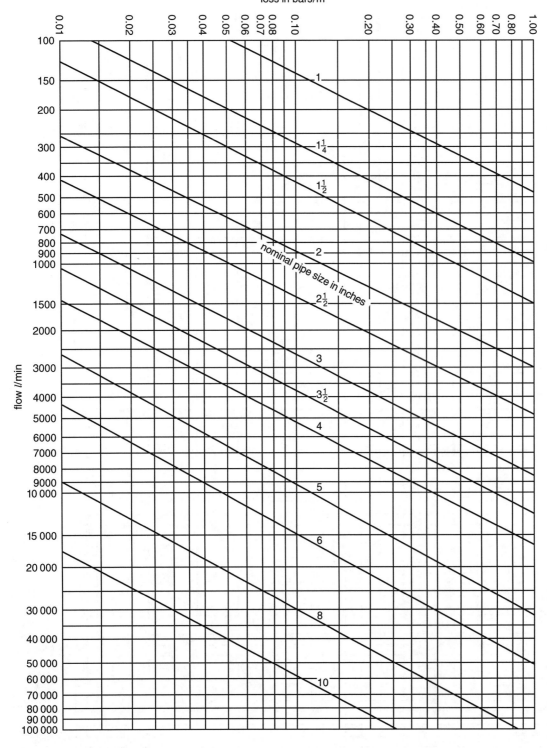

Multiply bars by 100 to get kPa.

[a] Use Table 14 to convert to other C-values.

Reprinted with permission from NFPA 15, *Water Spray Fixed Systems*, copyright © 1990, National Fire Protection Association, Quincy, MA 02269. This reprinted material is not the complete and official position of the National Fire Protection Association on the referred subject which is represented only by the standard in entirety.

APPENDIX D

Equivalent Lengths of Valves and Fittings
for Fire Protection Systems
(Hazen-Williams $C = 120$)
(all types of approved pipe)
(fittings and valves expressed in equivalent feet (m) of pipe)

	$\frac{3}{4}$ in	1 in	$1\frac{1}{4}$ in	$1\frac{1}{2}$ in	2 in	$2\frac{1}{2}$ in	3 in
45° elbow	1 (0.3)	1 (0.3)	1 (0.3)	2 (0.6)	2 (0.6)	3 (0.9)	3 (0.9)
90° standard elbow	2 (0.6)	2 (0.6)	3 (0.9)	4 (1.2)	5 (1.5)	6 (1.8)	7 (2.1)
90° long turn elbow	1 (0.3)	2 (0.6)	2 (0.6)	2 (0.6)	3 (0.9)	4 (1.2)	5 (1.5)
tee or cross (flow turned 90°)	4 (1.2)	5 (1.5)	6 (1.8)	8 (2.4)	10 (3.1)	12 (3.7)	15 (4.6)
gate valve	–	–	–	–	1 (0.3)	1 (0.3)	1 (0.3)
butterfly valve	–	–	–	–	6 (1.8)	7 (2.1)	10 (3.1)
swing check valve[a]	4 (1.2)	5 (1.5)	7 (2.1)	9 (2.7)	11 (3.4)	14 (4.3)	16 (4.9)

	$3\frac{1}{2}$ in	4 in	5 in	6 in	8 in	10 in	12 in
45° elbow	3 (0.9)	4 (1.2)	5 (1.5)	7 (2.1)	9 (2.7)	11 (3.4)	13 (4.0)
90° standard elbow	8 (2.4)	10 (3.1)	12 (3.7)	14 (4.3)	18 (5.5)	22 (6.7)	27 (8.2)
90° long turn elbow	5 (1.5)	6 (1.8)	8 (2.4)	9 (2.7)	13 (4.0)	16 (4.9)	18 (5.5)
tee or cross (flow turned 90°)	17 (5.2)	20 (6.1)	25 (7.6)	30 (9.2)	35 (10.7)	50 (15.3)	60 (18.3)
gate valve	1 (0.3)	2 (0.6)	2 (0.6)	3 (0.9)	4 (1.2)	5 (1.5)	6 (1.8)
butterfly valve	–	12 (3.7)	9 (2.7)	10 (3.1)	12 (3.7)	19 (5.8)	21 (6.4)
swing check valve[a]	19 (5.8)	22 (6.7)	27 (8.2)	32 (9.8)	45 (13.7)	55 (16.8)	65 (19.8)

[a] average values

APPENDIX E

Approximate Properties of
Selected Flammable Liquids and Gases[a]

name	flash point (°F)	explosive limits (% by volume) lower	explosive limits (% by volume) upper	ignition temperature (°F)	vapor density (air = 1.0)
acetone	0	2.6	13.0	869	2.0
acetylene	gas	2.5	100.0	581	0.9
acrolein	−15	2.8	31.0	455	1.9
acrylonitrile	32	3.0	17.0	898	1.8
ammonia	gas	15.0	28.0	1204	0.6
amyl acetate (nor)	76	1.0	7.5	680	4.5
amyl acetate (sec)	89	–	–	–	4.5
benzene (benzol)	12	1.3	7.1	1040	2.8
benzine	(see petroleum ether)				
butadiene (1,3)	gas	2.0	12.0	788	1.9
butane (nor)	gas	1.8	8.4	761	2.1
butane (iso)	gas	1.8	8.4	860	2.1
butyl acetate (nor)	72	1.4	15.0	797	4.0
butyl alcohol (nor)	84	1.4	18.2	650	2.6
butyl alcohol (iso)	82	1.7	11.0	800	2.6
butyl ether	(see dibutyl ether)				
carbon disulfide	−22	1.3	50.0	194	2.6
carbon monoxide	gas	12.5	74.0	1204	0.97
cyclohexane	−4	1.3	7.8	473	2.9
cyclopropane	gas	2.4	10.4	932	1.5
denatured alcohol (95%)	60	–	–	750	1.6
dibutyl ether (nor)	77	1.5	7.6	382	4.5
diethylene dioxide	54	2.0	22.0	509	3.0
diethyl ether	−49	1.9	36.0	320	2.6
dimethyl ether	gas	3.4	27.0	662	1.6
dioxane-p	(see diethylene dioxide)				
divinyl ether	−22	1.7	27.0	680	2.4
ethane	gas	3.0	12.5	959	1.0
ether	(see diethyl ether)				
ethyl alcohol	55	3.5	19.0	689	1.6
ethyl ether	(see diethyl ether)				
ethylene	gas	2.7	36.0	914	1.0
ethylene oxide	gas	3.0	100.0	804	1.5
formaldehyde	gas	7.0	73.0	806	1.1
gasoline (auto)	−50	1.3–1.4	6.0–7.6	700	3.0–4.0
heptane (nor)	25	1.1	6.7	419	3.5
heptane (iso)	< 0	1.0	6.0	536	3.5
hexane (nor)	−7	1.2	7.4	437	2.9
hexane (iso)	−20	1.0	7.0	–	3.0
hydrogen	gas	4.0	75.0	752	0.1
hydrogen sulfide	gas	4.3	45.0	500	1.2
isobutyl alcohol	(see butyl alcohol (iso))				
isobutane	(see butane (iso))				
isopentane	(see pentane (iso))				
isopropyl alcohol	(see propyl alcohol (iso))				

APPENDIX E

Approximate Properties of
Selected Flammable Liquids and Gases[a]
(continued)

name	flash point (°F)	explosive limits (% by volume)		ignition temperature (°F)	vapor density (air = 1.0)
		lower	upper		
methane	gas	5.0	15.0	1004	0.6
methyl alcohol	54	6.7	36.0	725	1.1
methyl ether	(see dimethyl ether)				
methyl ethyl ketone	28	1.9	10.0	960	2.5
naptha, VM&P	20–45	0.9	6.0	450–500	3.8
octane (nor)	56	1.0	3.0	428	3.9
octane (iso)	10	1.0	6.0	784	3.9
pentane (nor)	–57	1.4	7.8	500	2.5
pentane (iso)	–60	1.4	7.6	788	2.5
petroleum ether (benzine)	0	1.1	5.9	550	2.5
propane	gas	2.1	9.5	842	1.6
propyl alcohol (nor)	59	2.1	13.5	824	2.1
propyl alcohol (iso)	53	2.5	12.0	750	2.1
propylene	gas	2.4	11.0	860	1.5
toluene	40	1.3	7.0	896	3.1
vinyl acetate	18	2.6	13.4	800	3.0
vinyl chloride	gas	3.6	33.0	882	2.2
vinyl ether	(see divinyl ether)				
xylene (meta)	77	1.1	7.0	986	3.7
xylene (ortho)	63	1.0	6.0	869	3.7
xylene (para)	77	1.1	7.0	986	3.7

[a] The data in this table have been compiled primarily from *Flammability Characteristics of Combustible Gases and Vapors* by M.G. Zabetakis, U.S. Department of the Interior, Bureau of Mines Bulletin 627, Washington, D.C. 1965.

APPENDIX F
Standardized Sprinkler System Worksheet

Contract No. _____   Sheet No. _____ of _____

Name and Location _____

nozzle type and location	flow gpm (l/min)	pipe size in	fitting and devices	pipe equivalent length	friction loss psi/ft (bars/m)	required pressure psi (bars)	normal pressure	notes
$\frac{q}{Q}$			length			p_t	p_t	
			fittings			p_f	p_v	
			total			p_e	p_n	
$\frac{q}{Q}$			length			p_t	p_t	
			fittings			p_f	p_v	
			total			p_e	p_n	
$\frac{q}{Q}$			length			p_t	p_t	
			fittings			p_f	p_v	
			total			p_e	p_n	
$\frac{q}{Q}$			length			p_t	p_t	
			fittings			p_f	p_v	
			total			p_e	p_n	
$\frac{q}{Q}$			length			p_t	p_t	
			fittings			p_f	p_v	
			total			p_e	p_n	
$\frac{q}{Q}$			length			p_t	p_t	
			fittings			p_f	p_v	
			total			p_e	p_n	
$\frac{q}{Q}$			length			p_t	p_t	
			fittings			p_f	p_v	
			total			p_e	p_n	
$\frac{q}{Q}$			length			p_t	p_t	
			fittings			p_f	p_v	
			total			p_e	p_n	
$\frac{q}{Q}$			length			p_t	p_t	
			fittings			p_f	p_v	
			total			p_e	p_n	
$\frac{q}{Q}$			length			p_t	p_t	
			fittings			p_f	p_v	
			total			p_e	p_n	
$\frac{q}{Q}$			length			p_t	p_t	
			fittings			p_f	p_v	
			total			p_e	p_n	

APPENDIX G

Standard Symbols of Fire Sprinkler Components

sprinklers		piping, valves, control devices, hangers	
sprinkler, general	─○─	sprinkler piping and branch line	───
upright sprinkler	─○─	pipe hanger	─/─
pendent sprinkler	─○─	sprinkler riser	⊗
upright sprinkler, nippled up	─◎─	valve (general)	─▷◁─
pendent sprinkler, on drop nipple	─●─	angle valve (angle hose valve)	▷
sprinkler, with guard	─⊗─	check valve (general)	─N─ or ─→─
sidewall sprinkler	▽	alarm check valve	⧖●
outside sprinkler	▽	dry pipe valve	▷○ or ◆
special spray nozzle	─▲─	dry pipe valve with quick-opening device (accelerator or exhauster)	▷○
		deluge valve	◇
		OS & Y valve	┼
		alarm valve	─▶─
		preaction valve	─◁▷─
		fire department connection	─<

APPENDIX H

Internal Dimensions of Standard Steel Pipe

nominal pipe size	outside diameter		schedule 10[a]				schedule 30				schedule 40			
			inside diameter		wall thickness		inside diameter		wall thickness		inside diameter		wall thickness	
in	in	(mm)	in	(mm)	in	(mm)	in	(mm)	in	(mm)	in	(mm)	in	(mm)
1	1.315	(33.4)	1.097	(27.9)	0.109	(2.8)	–	–	–	–	1.049	(26.6)	0.133	(3.4)
$1\frac{1}{4}$	1.660	(42.2)	1.442	(36.6)	0.109	(2.8)	–	–	–	–	1.380	(35.1)	0.140	(3.6)
$1\frac{1}{2}$	1.900	(48.3)	1.682	(42.7)	0.109	(2.8)	–	–	–	–	1.610	(40.9)	0.145	(3.7)
2	2.375	(60.3)	2.157	(54.8)	0.109	(2.8)	–	–	–	–	2.067	(52.5)	0.154	(3.9)
$2\frac{1}{2}$	2.875	(73.0)	2.635	(66.9)	0.120	(3.0)	–	–	–	–	2.469	(62.7)	0.203	(5.2)
3	3.500	(88.9)	3.260	(82.8)	0.120	(3.0)	–	–	–	–	3.068	(77.9)	0.216	(5.5)
$3\frac{1}{2}$	4.000	(101.6)	3.760	(95.5)	0.120	(3.0)	–	–	–	–	3.548	(90.1)	0.226	(5.7)
4	4.500	(114.3)	4.260	(108.2)	0.120	(3.0)	–	–	–	–	4.026	(102.3)	0.237	(6.0)
5	5.563	(141.3)	5.295	(134.5)	0.134	(3.4)	–	–	–	–	5.047	(128.2)	0.258	(6.6)
6	6.625	(168.3)	6.357	(161.5)	0.134[b]	(3.4)	–	–	–	–	6.065	(154.1)	0.280	(7.1)
8	8.625	(219.1)	8.249	(209.5)	0.188[b]	(4.8)	8.071	(205.0)	0.277	(7.0)	–	–	–	–
10	10.75	(273.1)	10.37	(263.4)	0.188[b]	(4.8)	10.14	(257.6)	0.307	(7.8)	–	–	–	–

[a] Schedule 10 defined to 5 in (127 mm) nominal pipe size by ASTM A135.
[b] Wall thickness specified in NFPA 13, 2-3.2.

Reprinted with permission from NFPA 13, *Standard for the Installation of Sprinkler Systems*, copyright © 1991, National Fire Protection Association, Quincy, MA 02269.

APPENDIX I

Internal Dimensions of Standard Copper Tube

nominal tube size	outside diameter		type K				type L				type M			
			inside diameter		wall thickness		inside diameter		wall thickness		inside diameter		wall thickness	
in	in	(mm)	in	(mm)	in	(mm)	in	(mm)	in	(mm)	in	(mm)	in	(mm)
$\frac{3}{4}$	0.875	(22.2)	0.745	(18.9)	0.065	(1.7)	0.785	(19.9)	0.045	(1.1)	0.811	(20.6)	0.032	(0.8)
1	1.125	(28.6)	0.995	(25.3)	0.065	(1.7)	1.025	(26.0)	0.050	(1.3)	1.055	(26.8)	0.035	(0.9)
$1\frac{1}{4}$	1.375	(34.9)	1.245	(31.6)	0.065	(1.7)	1.265	(32.1)	0.055	(1.4)	1.291	(32.8)	0.042	(1.1)
$1\frac{1}{2}$	1.625	(41.3)	1.481	(37.6)	0.072	(1.8)	1.505	(38.2)	0.060	(1.5)	1.527	(38.8)	0.049	(1.2)
2	2.125	(54.0)	1.959	(49.8)	0.083	(2.1)	1.985	(50.4)	0.070	(1.8)	2.009	(51.0)	0.058	(1.5)
$2\frac{1}{2}$	2.625	(66.7)	2.435	(61.8)	0.095	(2.4)	2.465	(62.6)	0.080	(2.0)	2.495	(63.4)	0.065	(1.7)
3	3.125	(79.4)	2.907	(73.8)	0.109	(2.8)	2.945	(74.8)	0.090	(2.3)	2.981	(75.7)	0.072	(1.8)
$3\frac{1}{2}$	3.625	(92.1)	3.385	(86.0)	0.120	(3.0)	3.425	(87.0)	0.100	(2.5)	3.459	(87.9)	0.083	(2.1)
4	4.125	(104.8)	3.857	(98.0)	0.134	(3.4)	3.905	(99.2)	0.110	(2.8)	3.935	(99.9)	0.095	(2.4)
5	5.125	(130.2)	4.805	(122.0)	0.160	(4.1)	4.875	(123.8)	0.125	(3.2)	4.907	(124.6)	0.109	(2.8)
6	6.125	(155.6)	5.741	(145.8)	0.192	(4.9)	5.845	(148.5)	0.140	(3.6)	5.881	(149.4)	0.122	(3.1)
8	8.125	(206.4)	7.583	(192.6)	0.271	(6.9)	7.725	(196.2)	0.200	(5.1)	7.785	(197.7)	0.170	(4.3)
10	10.13	(257.3)	9.449	(240.0)	0.338	(8.6)	9.625	(244.5)	0.250	(6.4)	9.701	(246.4)	0.212	(5.4)

Reprinted with permission from NFPA 13, *Standard for the Installation of Sprinkler Systems*, copyright © 1991, National Fire Protection Association, Quincy, MA 02269.

INDEX

Underwriters Laboratories (UL), 9
Uniform Building Code, 1
Uniform Fire Code, 1
Upper explosive limit (UEL), 49
Upright sprinkler, 28
Urea-potassium bicarbonate, 15

Value, heating, 3
Valve, fast-acting gate, 53
Valve monitor switch, 11
Valves, types of, 22, 23, 25
Vapor density, 49
Velocity pressure, 37
Vent
   explosion, 50
   relief, 53
   relief, bursting-panel, 53
   roof, 5
Ventilation, dilution, 54
Venting explosion, 53
Vertical pump, 26
Vessel, steel, 34
Volumetric limits, 5

Wallboard, gypsum, 4
Water
   sources of, 24
   wet, 14
Water-flow detector, 11
Water hammer, 22
Water Spray Fixed Systems,
   NFPA 15, 4, 19
Water spray sprinkler system, 19
Water-supply valve, 22
Water tanks, types of, 25
Water tubing, 21
Weatherability, 13
Wet barrel, 26
Wet pipe sprinkler system, 17
Wet scrubber, 7
Wet standpipe, 15
Wet water, 14
Wiring, 12
Wood frame construction, 3
Worksheet, 42
Wrought steel pipe, 21

X-purging, 51

Y-purging, 51

Zoning, 12
Z-purging, 51